# Plant Image Analysis: Fundamentals and Applications

# Plant Image Analysis: Fundamentals and Applications

Editor

**Raj Patil**

scitus
academics

**Plant Image Analysis: Fundamentals and Applications**

Edited by **Raj Patil**

Printed in 2017

ISBN: 978-1-68117-020-6

Library of Congress Control Number: 2015931822



SCITUS Academics LLC,
616, Corporate Way, Suite 2, 4766,
Valley Cottage, NY 10989

www.scitusacademics.com



**Notice**

# Contents

biologically active substances. It bases on the direct observations of secretions released from allelopathic and medicinal species and the cell-donor interactions with cell-acceptors as biosensors (unicellular plant generative and vegetative microspores). Special attention was paid to the interactions with pigmented and fluorescing components of the secretions released by the cells-donors from plant species. Colored components of secretions are considered as histochemical dyes for the analysis of cellular mechanisms at the cell-cell contacts and modelling of cell-cell interactions. The fluorescence of plant biosensors was also recommended for the testing of natural plant excretions as medical drugs.

# INTRODUCTION

The visible (400–700 nm) fluorescence of intact living cells (usually called autofluorescence) excited by ultraviolet, violet, or blue light is often observed under luminescence microscope [1, 2]. Due to special microspectrofluorimeters or confocal microscope the fluorescence spectra of objects studied and the emission intensity can be also measured. Autofluorescence is the natural emission of cellular components due to a contribution of fluorescent compounds located in different cellular compartments. In general blue and green emission reflects the participation of universal cellular fluorophores NAD(P)H (max. 460 nm), pterins (max. 450 nm), and flavins (max. 520 nm) [3], although in dry state weak blue fluorescence (420–450 nm) is also peculiar to isolated nucleotides, RNA, and DNA as well as some proteins, and seen in the films under luminescence microscope [4, 5], but water quenches the emission [4]. Besides, there are characteristic fluorescent compounds in various organisms [6]. In animal cells chitin (max. 450–460 and 520 nm), collagen and elastin (max. 400–430, 465, 495 and 520 nm) or cellulose (max. 420–430 nm) in plants, and fungi can also contribute in blue fluorescence [6]. Orange-red fluorescence depends on organism. In plant cells main known fluorophore used for environmental monitoring is chlorophyll with maximum 675–680 nm [7] and in photosynthesizing bacteria: bacteriochlorophyll [6]. Uroporphyrinogen and haemoglobin contribute in red lightening of mammalians, while green-red range emission may reflect a contribution of sea animal fluorescent proteins used for the genetic experiments as vector for the study of cellular components [6–8]. Terpenes, some alkaloids (colchicine) and flavonoids in plants, emit in blue or in blue-green (470–525 nm) region of the visible spectrum whereas polyacetylenes, isoquinoline, and acridone alkaloids emit in yellow and orange [8]. Red fluorescence of plant cells may be also due to anthocyanins and azulenes (600–630 nm) [8]. At ageing or, if the object being undergone ozone and relative reactive oxygen species, fluorescent pigments lipofuscins are formed that may emit in wider spectral range—from blue (450–480 nm) to red (>600 nm) in a dependence on the nature of the organism and duration of the exposure under unfavourable factor [9, 10].

Autofluorescence of the cellular compounds may serve (1) as markers for cytodiagnostics and (2) considerations of as natural indicator of cellular state because the emission often shows changes in cellular metabolism and responses to the external and internal signals [8]. This is true for the plant secretory structures analyzed under luminescent microscope (the structures are not seen in usual microscope without special histochemical staining) or laser-scanning confocal microscope (LSCM images) [8, 10, 20–22, 37–39]. The intensity and spectral composition of the emission differ in a dependence on, namely, (1) nature of the organism or (2) cell analyzed, (3) a taxonomic position of the organism tested, and (4) environmental and experimental conditions (actinic light, temperature, humidity, physiological state of a cell as a whole and phase of development, influence of neighbour cells or organisms, including parasites). Autofluorescence could be used (i) in express-microanalysis of the accumulation of the secondary metabolites in secretory cells without long biochemical procedures; (ii) in diagnostics of cellular damage; (iii) in analysis of cell-cell interactions [8,10, 20, 21, 39].

In any case the phenomenon of autofluorescence could be used as a resource for the nondestructive vital microscopic analysis of natural organisms [41]. It has already practical application in various diagnostic procedures for cell biology analysis, ecology, medicine, and pharmacology. This paper considers the application of various types of luminescence microscopy to the study of plant cells, mainly such as secretory systems.

# IMAGES AND THE FLUORESCENCE SPECTRA OF FLUORESCING CELLS IN CYTODIAGNOSTICS

Today our knowledge in the problem of auto fluorescence, peculiar to living cells, mainly include results of experiments made on big plant cells that are suitable for observation both in vivo (such as in nature) and in vitro (in cultivated systems). Among similar objects are generative cells of flowers: pollen (male gametophyte) and pistil (female gametophyte) and non-generative cells, such as plant vegetative microspores and special secretory cells. Some examples dealt with the application of luminescence microscopy in various modifications to analysis of the cells are shown on Figure 1. In particular there are the images seen under usual luminescence or confocal microscopes and the fluorescence spectra measured by microspectrofluorimeter or some types of confocal microscopes. We shall also comment about Figure 1 in sections below.

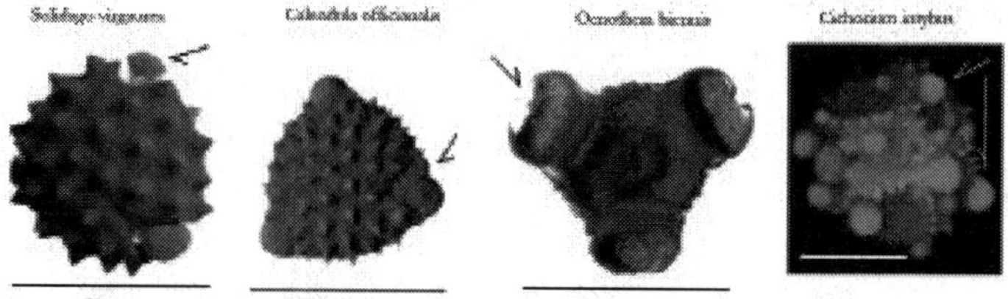

(a) Laser-scanning confocal microscopy of pollen

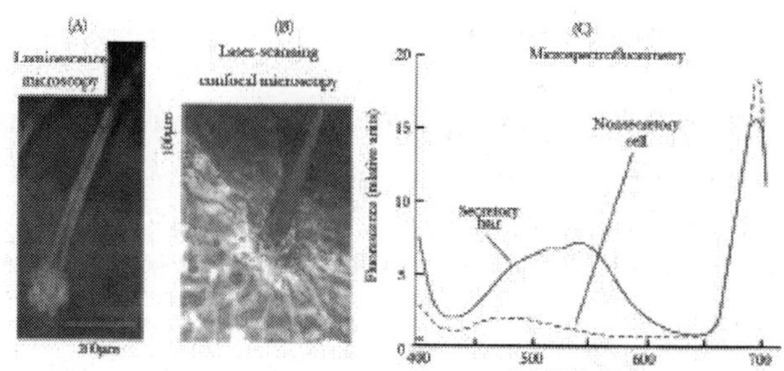


(b) Leaf secretory hair of symphytum officinale

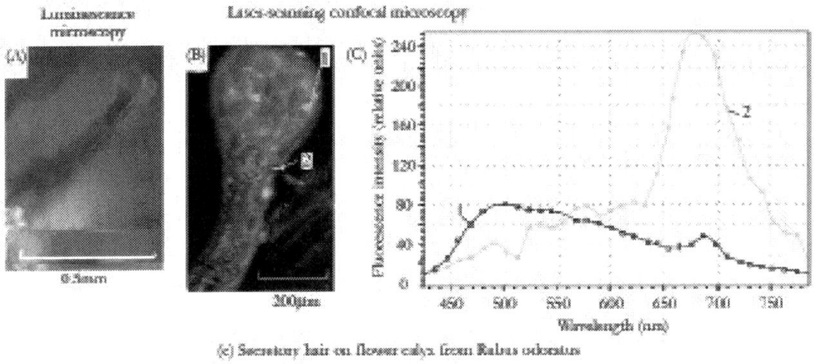


**Figure 1**: Images and the fluorescence spectra of some intact fluorescing plant secreting cells. Source: adapted from [21, 22] and unpublished data of the review' author. (a) LSCM images of pollen grains from various species under confocal microscope LSM 510 NLO "Carl Zeiss" (Laser 488 nm). Drops of secretions from apertures are seen. (b) and (c) Views of secretory hairs on leaf of comfrey Symphytum officinale and on the flower calyx of flowering raspberry Rubus odoratus under (A) luminescence microscope Leica 6000 (Excitation 360–380 nm) and (B) laser-scanning confocal microscope Leica TSC SP 5 (laser 405 nm) and the (C) fluorescence spectra measured by registered microspectrofluorimeter MSF-1 (Excitation 360–380 nm) or laser-scanning confocal microscope Leica TSC SP 5 from secretory head with released secretion (1) and stalk (2) of secretory hair.

The research of fluorescence maxima in secretory cells showed the correlations and differences in the emission spectra excited by UV-light among the plant species studied and their specialized structures [8]. For example, nectaries mainly fluoresce, in blue region, while sporangia of fungi, mosses, horsetails, and ferns are in green-yellow (500–550 nm). Except fungi, the structures have also maximum 680 nm in red, peculiar to chlorophyll located in chloroplasts. Crystals of secretions emit more brightness than liquids. Today the secretion peaks were compared with the maxima of individual components of the same secretions in solutions or films [5, 8,10]. Interference of various fluorophores in total emission may take place, but diagnostics of prevailing component of the secretion is real. Changes in the fluorescing compounds depend on stage of plant development, the influence of various factors—ozonation, redox agents and excretions of other organisms in the environment.

## Pollen

Pollen is the unicellular structure (served for plant breeding of seed-bearing plants) that germinates on pistil stigma of flower forming pollen tube that grows within pistil style and carries out two spermia—male cells. As soon as the pollen tube achieves ovule (seed bag), here one sperm interacts with egg cell, and the fertilization is over by the embryo arising. Pollen often show visible autofluorescence excited by ultra violet or violet light. Luminescence microscope or laser-scanning confocal microscope was used for both the receiving fluorescing cell images and (in some cases) fluorescence spectra. Some examples of the emission are on Figure 1(a) and Table 1. Full information about the earlier studies of pollen autofluorescence as well as apparatuses for the observation and measurement has been summarized in reviews [18, 19] and monograph [8]. Here there are the references related to pioneers who first have seen autofluorescence of pollen under a luminescence microscope [43–45] and later measured first fluorescence spectra of pollen from various species [46, 47] with a microspectrofluorimeter.

**Table 1**: Autofluorescence in pollen diagnostics (LM—Luminescence microscopy, MS—Microspectrofluorimetry, LSCM—Laser-scanning confocal microscope)

| Diagnostics | Microscopic method | Fluorescence characteristics |
|---|---|---|
| Vital pollen characteristics. Changes in the autofluorescence of the pollen wall during microsporogenesis and chemical treatments [11, 12] | LM, MS | Color and fluorescence spectra |
| Wall development of sterile and fertile pollen [13] | LM | Color |
| Assessment of pollen retention using exine autofluorescence [14] | LM | Images |
| Matured and non-matured pollen [15–19] | MS | Fluorescence spectra (pigment composition) |
| Self-incompatibility and self-compatibility of pollen [17] | MS | Fluorescence spectra (pigment composition) |
| Pollen loads analysis as well as pollen-derived products of bees [8, 17] | MS | Fluorescence spectra (pigment composition—azulenes, carotenoids) |

| | | |
|---|---|---|
| Pollen-pollen interactions in the mixtures [8, 20–23] | LSCM, MS | Changes in the fluorescence spectra of pollen from different species reflect changes in pigments' state |
| Pollen viability [8, 17] | MS | Fluorescence intensity after moistening |
| Pollen-pistil interaction at fertilization [8, 17, 24] | MS | Changes in the fluorescence spectra reflect the receptor state |
| Exine structure [25] | LSCM | Images |
| Quantifying dynamics of pollen shed [26] | LM | Images |
| Ozone damages [27–29] | MS | Changes in the fluorescence spectra reflect pigments' state and forming pigment of aging lipofuscin |
| Damages by reactive oxygen species [28, 29] | MS | Changes in the fluorescence spectra |
| Classification of pollen grain species using autofluorescence image analysis [30, 31] | LM | Images and blue/red autofluorescence ratio |
| Testing in nanotechnology: Commercially produced fullerene fine particles [32] | LM | Adhesion of nanoparticles estimated on the autofluorescence intensity changes |
| The search of fruitful varieties from certain plant species [33] | LSCM | Images |
| Genetic screen to identify genes involved in pollen exine production [34] | LSCM | Wild-type exine fluoresces with two peaks 460 nm (higher) and 505 nm (lower) |
| A novel screen identifies multiple pollen mutants that have defective exine [34] | LSCM | Defective mutants differ in the emission intensity in the above-mentioned peaks |
| Taxonomic classification of pollen [35] | LSCM | Images |
| Pollen concentration in air [6, 36] | LM, LSCM | Images, emission intensity |

Figure 1 shows the LSCM images of pollens (from medicinal plants) released fluorescing secretions from apertures, through which pollen tube enters at pollen germination. We see different colors of autofluorescence from pollen surface and drops of secretions (marked with arrows) released from apertures (green emission of the secretions from Solidago virgaurea, orange-red lightening from one of Calendula officinalis, and colourless from Oenothera biennis). Today the phenomenon of autofluorescence is applied to the various diagnostics seen from Table 1. First papers in this line dealt with analysis of the exine autofluorescence by luminescence microscope. Exine, the outer layer of pollen cover, consists of sporopollenin and in various species may include various fluorescent pigments from phenols and carotenoids up to azulenes and proazulenes [8, 15]. Structure of exine was also analyzed with confocal microscopy [25], permitting to receive optical slices [8, 20, 21, 38, 39], and the fluorescence spectra were also recorded [8, 15–19]. Based on the autofluorescence the various steps in pollen development were studied [8, 13]. Moreover, the phenomenon is applied to pollen classification [30,31], including the search pollen grains of fruitful varieties from certain plant species [33]. For understanding of sporopollenin synthesis and exine assembly a simple effective genetic screen based on their autofluorescence characteristics was represented for the Arabidopsis thaliana pollen (the recovery of mutants with a variety of defects in exine structure) to identify genes involved in pollen exine production [34].

The fluorescence spectra of pollen from 20 plant species, both wind—and insect-pollinated, were analyzed [8]. Pollen grains from wind-pollinated species (birch, pine and others) have mainly blue fluorescence whereas insect-pollinated ones belonged to different species—in blue, yellow, green, or blue-green. Nonmatured pollen demonstrated red fluorescence due to the presence of chlorophyll, while matured grains changed the color of the emission. The autofluorescence spectra of many pigmented and colorless pollens treated with ozone in various concentrations was also studied, and some characteristic damages were shown [27–29]. The fluorescence maximum at 530–550 nm disappeared in carotenoid-containing pollen grains of Philadelphus grandiflorus and Epiphyllum hybridum, and a new maximum at 475–480 nm arose (unlike carotenoid-less pollen of Plantago major) that correlated with lipofuscin-like substances observed in the extracts from the pollen grains [27–29].

In pollen mixtures changes in the fluorescence of pollens from different species may reflect their chemical (allelopathic) relations [8, 21]. LSCM images directly demonstrated the alterations in the emission and the state of pollen grains from Field Scabious Knautia arvensis (L.) Coult. (Dipsacaceae) contacting with pollen from wormwood Artemisia absinthium L. (Linn.) (Asteraceae) [20].

The autofluorescence imaging alone or in a combination with other morphological parameters [33] can be considered as nondisruptive method for pollen classification since it does not require any treatment of fixing or staining of the specimens. Therefore, this simple and fast method can be used for pollen counting and taxonomical classification in multiple applications that concern disciplines like aerobiology for identification of pollens in bioaerosols [6, 36, 48, 49]. Today the pollen study is very important, because air-occurring pollen from certain plant species may induce allergy, although on the other side pollen grains of medicinal plants are used for drug preparation. Knowledge of pollen state is necessary for analysis of plant fertility in genetics and biotechnology or of the meteorological situation in ecological monitoring as well as in pharmacy.

## Pistil

Pistil of flower is female gametophyte which stigma receives chemosignal from pollen as male gametophyte and recognizes pollen grains from the same plant species (as well as self-compatible or self-incompatible pollen of the species) or from foreign species. Autofluorescence reflects the pistil state before pollen application and after the pollen grains addition on pistil stigma [8, 17, 24]. If self-pollination occurs one can see changes in the peaks positions and the emission intensity in a comparison with the pistil fluorescence reply on pollen that originates from another plant of the same species like was seen on the example with Hippeastrum hybridum[16, 17]. When foreign pollen was applied to the pistil stigma often there are no changes in the pistil fluorescence [8]. Moreover pistils from self-compatible and self-incompatible lines of the one and same plant species, in particular as in Petunia hybrida clones, have different maxima in the fluorescence spectra [8].

## Secretory Cells—Donors of Biologically Active Fluorescing Compounds

Any living cells may form and release secretions out, and it is better seen in specialized secretory plant cells with fluorescent products [8, 10]. Images of the secreting cell fluorescence excited with ultraviolet or violet light are especially valuable in allelopathic and medicinal plants

## The Parasitive Invasion

Figure 2 demonstrates the example for leaf of medical plant mint Mentha piperita used by aphid for nutrition. On the leaf colorless microscopic larvae are plant-louse not seen well up to the excitation by actinic light of luminescence microscope. Blue-fluorescing Mentha gland (if excited by UV-light) with maxima 475, 540, 675–680 nm in the fluorescence spectrum is covered by weak blue-fluorescing aphis larva (chitin emission, mainly at 450 nm). At excitation by violet or green-yellow light yellow-fluoresced mint gland (surrounded by red fluorescing non-secretory cells of leaf mesophyll due to chlorophyll presence) mainly was seen through nonvisible (transparent) larva body.

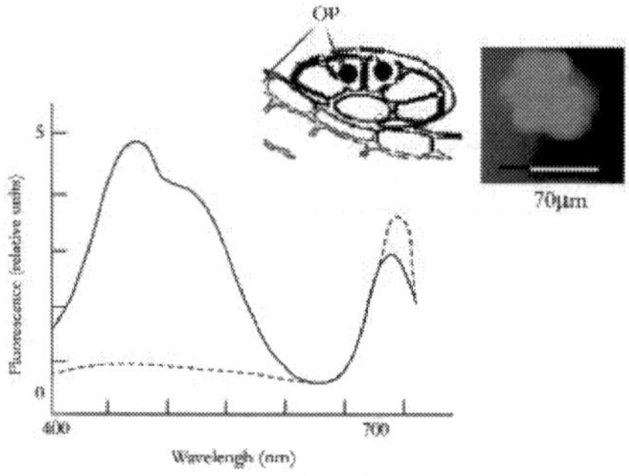


(a)

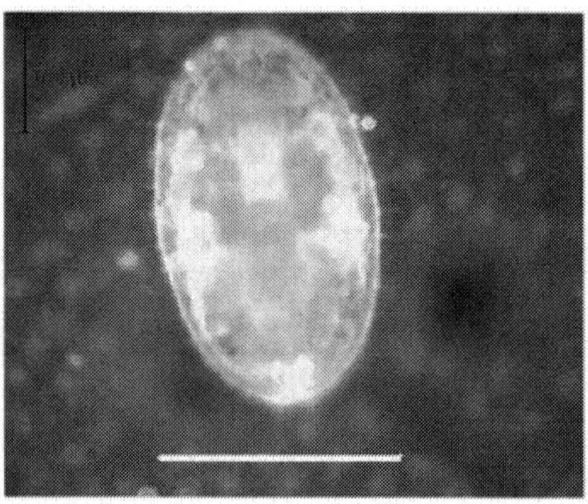


(b)

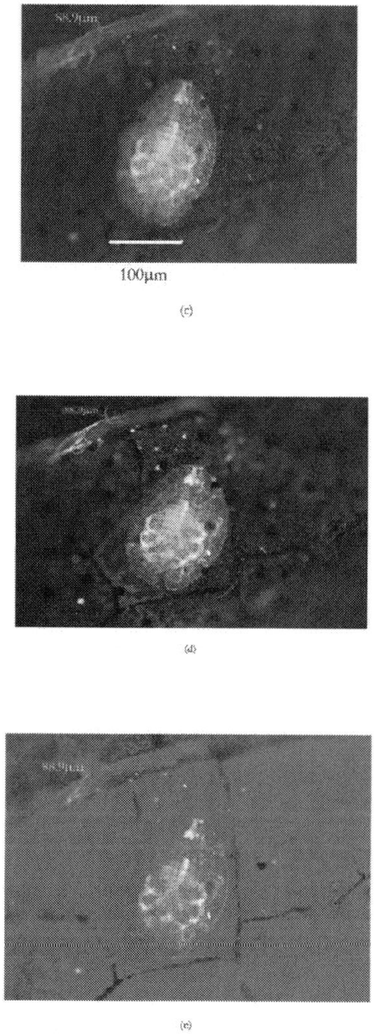


**Figure 2**: Fluorescent larva of aphis interacts with cellular complex of leaf gland of mintMentha piperita. (a) Left: the fluorescence spectra of secretory cells measured by microspectrofluorimeter MSF-1. Excitation: 360–380 nm. Unbroken line—secretory cells; broken line nonsecretory cells. The position of optical probe (OP) is shown. Right: the common view of gland under luminescence microscope "Carl Zeiss CD-28". ((b)–(e)) Images under luminescence microscope Leica DM 6000. (b) The common view of aphis larva out of the gland, ((c)–(e)) views of larva on the mint gland. Excitation by light 360–410 nm (b), 420–450 nm (c), 480 nm (d), and 560 nm (e).

Many pathogenic and saprophytic fungi fluoresce in blue green (in 390–490 nm spectral region) under actinic ultraviolet light [54–57]. Moreover the phenomenon was better seen under stress (desiccation, heat, chemicals) [56]. Recently blue-green autofluorescence of fungi of Ustilago maydis on maize and cell wall of this plant is an indicator of the invasion [57]. At the fungal infection the value of autofluorescence as a screening method for detecting fungi in tissues was compared with conventional histochemical stains. Autofluorescence can be used as a rapid screening method for identification of fungi in tissue sections as it does not require any other specialized staining procedure. Confocal microscopy appears to be more suitable for plant-pathogen interaction in details [58]. Fungal fluorophores may depend on species and are not well studied yet. Today fungal infection is better registered observing the enhanced blue-green autofluorescence of plant cell walls related to their phenols and sometimes auxin exchange [59].

## Pollen-Pollen Interactions

Special interest is the visualization of the interaction between pollen grains of various species that named as pollen allelopathy. At the contact, excretion from pollen-donor acts as chemosignal on pollen-acceptor, inducing the changes in the fluorescence of the microspores and their germination. The interaction of big pollen from Knautia arvensis (fluorescing in green) and small pollen grains from Artemisia absinthium or from Solidago virgaurea, which have released drops of red-fluorescing secretion as seen under laser-scanning confocal microscope [8, 20, 21]. This secretion was seen on the surface of non-germinated pollen from Knautia arvensis and also on the surface of pollen tubes. The interaction of green fluorescent pollen of K. arvensis and red-fluorescing secretion of pollen from Artemisia absinthium decreased the germination in both contacting types of microspores. In this case, perhaps, the enhancement in fluorescence is related to the energy loss peculiar to nonviable pollens. On LSCM images of fluorescent pollens in their mixtures red-fluorescing secretion for large pollen of Oenothera biennis contacted with small pollens of Matricaria chamomilla emitted in green and liberated green-fluorescing secretions were seen [21]. Optical slices showed the interior of contacted pollen grains, green fluorescence was seen only on the surface of M. chamolilla microspore, while red emission was also seen inside the pollen grain of O. biennis. LSCM images of fluorescent pollen made it possible to observe directly (i) the contact of intact cell, (ii) changes in their emission, and (iii) location of contacting secretions.

Intensity of autofluorescence from interacting cells in pollen mixtures also reflects the mutual influence of contacted secretions as shown for microspores from Hippeastrum hybridum, Philadelphus grandiflorus,Knautia arvensis, Plantago major, and Oenothera biennis used as pollen-acceptors, while other pollen grains from 42 plant species—as donor pollens [21]. Change of the light emission (excited by UV 360–380 nm or violet 430 nm light) is a visual chemical response to the pollen excretion from foreign pollens. To develop models for pollen-pollen interactions, the effects of allelochemicals secreted by foreign pollen (i.e., donor pollen) were studied in vitro or in vivo on autofluorescence concerned with the germination of acceptor-pollen [21]. Autofluorescence of acceptor and donor pollens in mixtures was changed to blue-green and red spectral regions that may serve as the indicator reply on foreign microspores. Generally increased blue-green and occasionally red emission indicated the decrease in the rate of pollen germination. However, red fluorescence was not related to germination and depended, mainly, on species used. It may be explained by redox changes in pollen components (often cover constituents) or by the synthesis of new fluorescing metabolite. Perhaps, the enhancement in blue-green emission of pollen could be related to the formation of fluoresced Schiff bases or to the formation of NAD (P) H emitted with maxima 460 nm. Red fluorescence of matured pollen is often correlated with the presence of anthocyanins or azulenes like occurs in overgrown parts of plants [8, 21].

# AUTOFLUORESCENCE AS BIOSIGNAL OR MARKER TO STUDY LOCALIZATION OF SECONDARY METABOLITES

Even 50 years ago Frey-Wyssling and Agthe have discovered fluorescence of flower nectar [8], as possible signal for insects-pollinators [60], and today the phenomenon has been demonstrated for various floral and

their high activity and stability [71]. Criteria in the screening of the plant cells as biosensors for biologically active compounds should be as follows: (1) the object use is easy in laboratory and in the field and (2) autofluorescence should be one of the physiological responses on the external chemicals because reflects the fast biochemical changes. They could serve as the biosensors of some biologically active excretions (having drugs' characteristics) from plants known as weeds, medicinal, or poisonous plant species [37, 42, 71]. Plant biosensors are more suitable for similar testing, than animals, which use needs procedures of vivisection [40, 71].

## Characteristics of Biosensors

As unicellular biosensors vegetative microspores of spore-bearing plants and pollens of seed-bearing plants [8,42, 71] may be used as well as multicellular biosensor—pistil stigma in flowers [8, 24].

Figure 3 shows examples of fluorescent analysis of vegetative microspores of horsetail Equisetum arvense using various technique. Under luminescence microscope, green fluorescence of cell wall and elaters (shell employees to attach microspore to the substrate) is visible. Through the rather transparent cover bright red fluorescence of chloroplasts, located inside the cell among nonlightening cytoplasm, is also observed. The total spectra of fluorescence from cells measured by microspectrofluorimeter depended on the presence of appropriate pigments and had the three peaks—in blue with maximum 450–460 nm (azulenes, phenols), in green with maximum 540–550 nm (carotenoids), as well as the most intense red emission with maximum 675–680 nm (chlorophyll). Cell cover liberated during the microspore development does not have a maximum in red spectral region [72]. Average fluorescence intensity of 100 cells measured on the one subject slide by dual-beam microspectrofluorimeter in green (520 nm) and red (640–680 nm) was differed in many times as seen from numerical data of summary emission shown in the histogram area (3.15 relative units in red against 0.13 relative units in green region). Red emission prevailed. Its intensity increases during the microspore development in comparison with dry rest (nondeveloped) cells [72, 73].

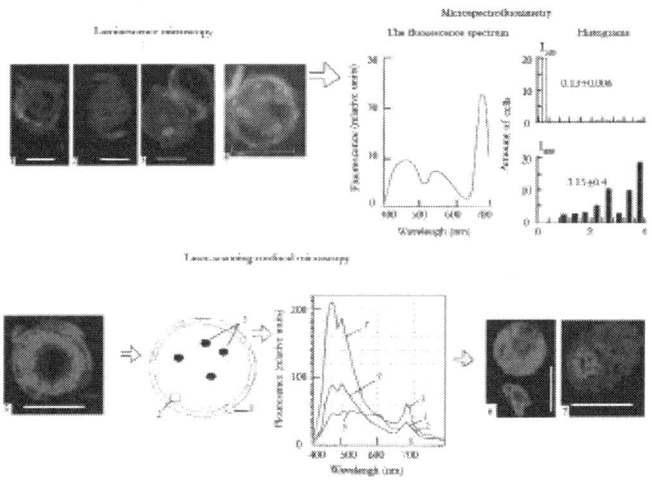

**Figure 3**: The images and fluorescence spectra of the vegetative microspores of horsetailEquisetum arvense. Sources: [8, 42] and nonpublished data of author. Upper side: Luminescence microscopy (left). Views (1–4) under the luminescence microscope Leica 6000. Microspectrofluorimetry. The fluorescence spectrum (middle) and the emission intensity distribution in form of numerical data or histogram (right). The fluorescence intensity (I) on histograms measured by dual-wavelength microspectrofluorimeter MSF-2 shows as $I_{520}$ and $I_{680}$, respectively, the emission intensity in the green region with maximum 520 nm and red region with a maximum 680 nm. Lower side: Laser-scanning confocal microscopy. Images 5–7 under laser-scanning confocal microscope LSM 510 NLO (5–moistured spore) and Carl Zeiss (6 and 7 developing spore that put off the blue-fluoresced rigid cover-envelope and then divided, resp.). On the scheme of optical image slice under laser-scanning confocal microscope (at the middle of the picture numbers mean cellular parts, from which the appropriate fluorescence spectra have been received with laser-scanning confocal microscope Leica TCS SP-5. By numerals exine (1), its inner layer intine with plasmic membrane (2) and chloroplast (3) are marked. Excitation for the luminescence microscope or microspectrofluorimeter was 420 nm and for laser-scanning confocal microscopy—405 nm laser.

If the division of the spore cell occurs (and duplication of chloroplasts takes place) the 2 times increase in red fluorescence is the indicator of cellular division because amounts of chloroplasts and fluorescing chlorophyll are doubled. Unlike the microspectrofluorimetry laser-scanning confocal microscopy permits to see images of optical slices of intact microspore cells at different depths and register the spectra of fluorescence from individual parts (marked as numerals on Figure 3), namely, cell wall, including its outer layer exine (1), its inner layer intine with plasmalemma (2), and chloroplasts (3).

The natural fluorescence of pollen and pistils as biosensors is served to study of various processes [10, 21, 22,24, 27–29] from the damages by ozone and other reactive oxygen species to chemosignaling. Indicators of the emission (Table 3) based on color and intensity of fluorescence altered under ozone and reactive oxygen species or the addition of some natural compounds (including complex secretions) and artificial chemicals. Pollen grains of Hippeastrum hybridum were often used as biosensors [37, 42, 71] as well as a pistil stigma of the same species [8, 24]. Multicolor pictures received looked as fantastic art photo that makes them fine illustrations [20, 21]. Besides fluorescent pollens, weak-emitted pollen such as from plantain Plantago major is also used for the analysis of interactions with fluorescent plant secretions, in particular from colored pollen grains) and the action mechanisms of drugs [8, 20–22, 40].

**Table 3**: Pollen biosensors

| Plant species | Colour of pollen autofluorescence | Prevailing fluorophore | Indicator reaction |
|---|---|---|---|
| Hippeastrum hybridum L. (Herb), knight's star (Amaryllidaceae) | Blue-green | Flavins, phenols | Shifts in the fluorescence maxima and intensity at pollen-pistil or pollen-pollen interactions and neurotransmitters additions [8, 16, 17,19, 21, 23] |
| Philadelphus grandiflorus, wild mock-orange (Hydrangeaceae) | Yellow-orange | Carotenoids, azulenes, anthocyanins | Shifts in the fluorescence maxima and intensity after ozone or other reactive oxygen species treatment [26–28] |
| Knautia arvensis (L.) Coult., field scabious (Dipsacaceae) | Green | Anthocyanins | Shifts in the fluorescence intensity at pollen-pollen interactions [20,21] |

## Biosensors in the Study of Chemosignal Processes

Chemosignaling is a first process occurred at contacts of sexual and somatic cells. For the studies it is necessary to distinguish two groups of fluorescing objects—cells which serve as donors of fluorescing compounds excreted and cells-acceptors of the chemical signal that are biosensors. Analyzing both cell types the researcher can observe, namely: (1) the changes in natural fluorescence of contacting cells, (2) the penetration of emitted fluorophore into the cell acceptor, (3) binding of the lightening compounds within cells, and (4) quenching or strengthening the total cell-acceptor fluorescence.

Chemosignaling at direct cell-cell contacts is possible to observe, using some sensitive pollen as biosensor (like to study of pollen allelopathy in Section 2.1). Application of LSCM technique permits to observe direct interaction of the contacted pollens and the alterations in their state, basing on the fluorescence changes. In example with LSCM images the secretion released from red-fluorescing pollen of Hypericum perforatum may stain the surface of the color-less pollen from Plantago major [8, 20]. The fluorescence of slices showed that the red-emitting pigment hypericin concentrated only on the surface without penetration into a cell.

Besides the direct observation of chemosignaling at cellular communications on LSCM-images, there is modelling of the interaction in the system biosensor–contacting chemical. Use of cellular autofluorescence as a response to the chemosignal can be one of the approaches to study mechanisms of action of biologically active substances on sensitive cells [37, 59]. Cellular mechanisms of the effects of many natural compounds as chemosignals often are not suitable to observe on the tissue slices of animals after vivisection or special conditions of cell storing and cultivation; plant cells have advantages as biosensors. For example, neurotransmitters acetylcholine, catecholamines, serotonin and histamine that also found in excretions of plant and microorganisms [74, 75] changed the position of maxima in the emission spectra and intensity of pollen and pistil autofluorescence in Hippeastrum hybridum [8, 24] or vegetative microspores of horsetailEquisetum arvense [42]. The neurotransmitters in the presence of other components of secretions—lipids and proteins or membrane-active agents of fungal infection—decreased or enhanced the fluorescence intensity at 680 nm of biosensors-vegetative microspores from horsetail Equisetum arvense [42]. Some of the contacted secretory products also fluoresce that allows observing how they interact with unicellular biosensors [8, 21, 22,24]. Fluorescent components of plant secretions are used to study of chemosignal processes at cellular level, stain living cells and provide valuable information about the level of their entering into a cell and binding sites [21, 22] as well as the mechanisms of chemoreception and action of some drugs [40, 42, 71]. Blue-fluorescent antagonists (blockers of appropriate receptors), in particular antagonist of acetylcholine atropine or d-Tubocurarine and antagonist of catecholamines yohimbine, were binding with plasmatic membrane while histaminic antagonist azulene penetrated into a cell that was seen on LSCM images [42]. Table 4 demonstrates the targets of the compounds' bindings. Antagonists of neurotransmitters are included in secretions of medicinal (allelopathically active) plants as pharmaceuticals with own fluorescence. All compounds studied decreased number of the cells of E. arvense fluorescing in red (Figure 4). d-Tubocurarine and yohimbine did not penetrate the cells binding on the cellular surface, but azulene was linked at both the cell surface and inside the cell, demonstrating blue or blue-green emission. Tubocurarine bound with cholinoreceptor in plasmatic membrane stimulated blue emission and decreased red chlorophyll fluorescence masking plastidal emission. The evidence that tubocurarine and acetylcholine are linked in one and the same site of the membrane is the absence or small blue fluorescence and enhanced red fluorescence in the variant with preliminary treatment of the biosensor cells with acetylcholine.

**Table 4**: Fluorescence of biologically active compounds (medical drugs) out and within cells of microspores—biosensors. Source: [38, 40]

| **Component of plant secretions with pharmaceutical features** | **Fluorescence (color/maxima in nm)** | | | **Cellular component—target of a fluorescent staining** |
|---|---|---|---|---|
| | Pure compound | At the interaction with microspores | | |
| | | E. arvense | H. hybridum | |
| Azulene | Blue/425 | Blue/425, 550 | Blue/425, 550 | Nucleus, chloroplasts |
| Atropine | Blue/415 | Blue/465–470 | Blue/465–470 | ChR in Plm |
| d-Tubocurarine | Blue/415 | Blue/425 | Blue/465–475 | ChR, Plm |
| Yohimbine | Blue/455 | Blue/4305 | Blue/475 | AdR, Plm |
| | | | | |

*Plm: plasmalemma, CW: cell wall, ChE: cholinesterase, ChR: cholinoreceptor, AdR: adrenoreceptor, and Glc: glucosidase.

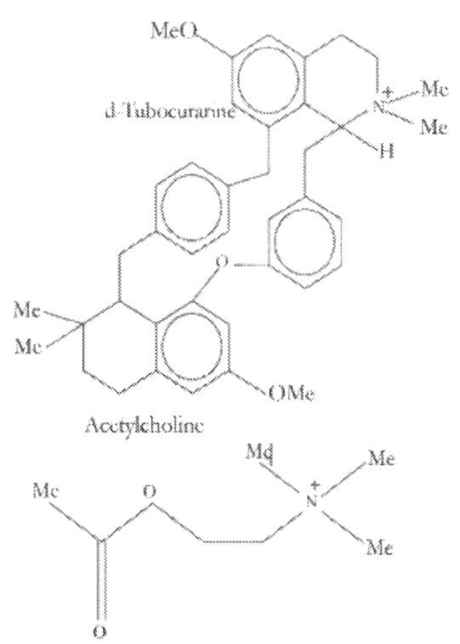


(a)

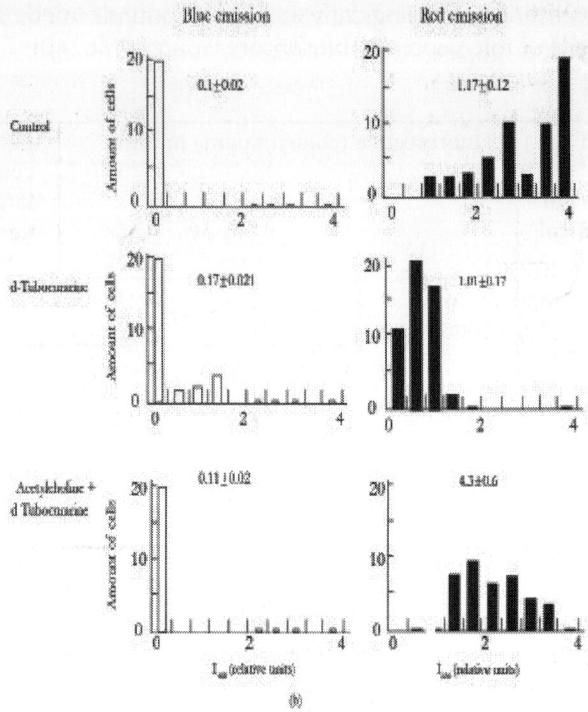


**Figure 4**: Summed distribution histograms of blue (maximum 460 nm) fluorescence intensity of vegetative microspores of horsetail Equisetum arvense before and after the treatment with components of plant secretions during 24 h. Excitation 360–380 nm. Total amount of studied cells in a variant = 100. Summed values for 100 analyzed cells are also given.

Besides unicellular microspores, the long-lengthed pistil of Hippeastrum hybridum which the stigma autofluorescence (maxima in the fluorescence spectra and the emission intensity) is sensitive to acetylcholine and histamine as well as their antagonists d-Tubocurarine and azulene [8, 24] may be suitable biosensor for the study of chemosignalling.

## Potentials for a Search of New Biosensors

Among potential biosensors may be multicellular systems having autofluorescence. Similar biosensor based on petal fluorescence in flowers such as Mirabilis jalapa may be also applied to analysis of chemosignalling dealt with certain pigments-fluorophores [60]. The visible fluorescence emitted by one pigment, a yellow betaxanthin, is absorbed by another, a violet betacyanin, to create a contrasting fluorescent pattern on the flower's petals for pollinators. The analysis of free radical processes with the participation of the pigments [76] opened new horizons to search biosensors among pigmented plant cells.

Perspective biosensors today appears to be mycorrhizal roots of plants sensitive to a colonization of arbuscular mycorrhizal fungal structures (such as hyphae, vesicles and spores) that brightly autofluorescence among root tissues under ultraviolet, blue or green light excitation [76–78]. Whole roots of lucerne Medicago sativa may be suitable for the same aim.

# CONCLUSION

Autofluorescence of living cells may serve as biosensor and bioindicator reaction for various studies of plants–from taxonomic position of plant or pollen viability to the chemosignalling processes. This reaction is analyzed by following ways, namely: (1) on undamaged structures, if their surface composed of fluorescent components or interior of the structures filled of fluorescent secretion; (2) modeling with individual cellular components showing the picture of the fluorescing cellular structure observed. The autofluorescent structure images are studied in a dependence on a purpose either for anatomic characteristics applying to taxonomic sign in plant systematics or for a comparison of samples in the control variants and those that exposed under someone factor. Moreover, optical slices made with laser-scanning confocal microscopy permit to show a location of fluorescing secondary metabolite naturally synthesized within the cell or exometabolite interacted with the cell at cell-cell communications or at experimental addition of the chemical agent. Besides images, the fluorescence spectra of emitted structures may be recorded with microspectrofluorimetry or certain laser-scanning confocal microscopes. Measuring the absorbance and fluorescence spectra there is a possibility to estimate quantitative yield of fluorescence not only for individual fluorophore, but also for certain cells, such as flower petals.

# REFERENCES

1. D. L. Taylor and E. D. Salmon, "Basic fluorescence microscopy," in Methods in Cell Biology: Living Cell in Culture, J. L. Wang and D. L. Taylor, Eds., pp. 207–237, Academic Press, San Diego, Calif, USA, 1989.
2. H. Andersson, T. Baechi, M. Hoechl, and C. Richter, "Autofluorescence of living cells," Journal of Microscopy, vol. 191, no. 1, pp. 1–7, 1998.
3. O. S. Wolfbeis, "The fluorescence of organic natural products," in Molecular Luminescence Spectroscopy: Methods and Applications, S. G. Schulman, Ed., pp. 167–370, John Wiley & Sons, New York, NY, USA, 1985.
4. V. V. Roshchina, "Cellular models to study the allelopathic mechanisms," Allelopathy Journal, vol. 13, no. 1, pp. 3–16, 2004.
5. V. V. Roshchina, V. A. Yashin, A. V. Yashina, and M. V. Gol'tyaev, "Colored allelochemicals in modelling of cell-cell allelopathic interactions," Allelopathy Journal, vol. 28, no. 1, pp. 1–12, 2011.
6. C. Pöhlker, J. A. Huffman, and U. Pöschl, "Autofluorescence of atmospheric bioaerosols—fluorescent biomoleculesand potential interferences," Atmospheric Measurement Techniques, vol. 4, no. 5, pp. 5857–5933, 2011.
7. G. Agati, "Response of the in vivo chlorophyll fluorescence spectrum to environmental factors and laser excitation wavelength," Pure and Applied Optics, vol. 7, no. 4, pp. 797–807, 1998.
8. V. V. Roshchina, Fluorescing World of Plant Secreting Cells, Science, Enfield, UK, 2008.
9. M. N. Merzlyak, "Liposoluble fluorescent "aging pigments" in plants," in Lipofuscin-1987: State of the Art, I. Nagy, Ed., pp. 451–452, Akademiai Kiado, Elsevier, Budapest, Amsterdam, The Netherlands, 1988.
10. V. V. Roshchina, "Autofluorescence of plant secreting cells as a biosensor and bioindicator reaction,"Journal of Fluorescence, vol. 13, no. 5, pp. 403–420, 2003.

11. M. T. M. Willemse, "Autofluorescence of pollen wall of Lilium and changes in pollen wall of Gasteriaand Lilium anther," Acta Societatis Botanicorum Poloniae, vol. 50, pp. 103–110, 1981.
12. M. N. B. M. Driessen, M. T. M. Willemse, and J. A. G. van Luijin, "Grass pollen grain determination by light- and UV-microscopy," Grana, vol. 28, no. 2, pp. 115–122, 1989.
13. J. C. Audran and M. T. M. Willemse, "Wall development and its autofluorescence of sterile and fertileVicia faba L. Pollen," Protoplasma, vol. 110, no. 2, pp. 106–111, 1982.
14. S. A. Weinbaum and V. S. Polito, "Assessment of pollen retention using exine autofluorescence,"Mikroskopie, vol. 42, pp. 278–280, 1985.
15. V. V. Roshchina, E. V. Melnikova, N. A. Spiridonov, and L. V. Kovaleva, "Azulenes, the blue pigments of pollen," Doklady Biological Sciences, vol. 340, no. 1, pp. 93–96, 1995.
16. V. V. Roshchina, E. V. Mel'nikova, and L. V. Kovaleva, "Autoflourescence in pollen-pistil system inHippeastrum hybridum," Doklady Akademii Nauk, vol. 349, no. 1, pp. 118–120, 1996.
17. V. V. Roshchina, E. V. Mel'nikova, and L. V. Kovaleva, "Changes in fluorescence during development of the male gametophyte," Russian Journal of Plant Physiology, vol. 44, no. 1, pp. 36–44, 1997.
18. V. V. Roshchina, E. V. Melnikova, L. V. Mit'kovskaya, and V. N. Karnaukhov, "Microspectrofluorimetry for the study of intact plant secreting cells," Zhurnal Obshchei Biologii, vol. 59, no. 5, pp. 531–554, 1998.
19. V. V. Roshchina and E. V. Melnikova, "Microspectrofluorimetry of intact secreting cells, with applications to the study of allelopathy," in Principles and Practices in Plant Ecology: Allelochemical Interactions, Inderjit, K. M. M. Dakshini, and C. L. Foy, Eds., pp. 99–126, CRC Press, Boca Raton, Fla, USA, 1999.
20. V. V. Roshchina, A. V. Yashina, and V. A. Yashin, "Cell communication in pollen allelopathy analyzed with laser-scanning confocal microscopy," Allelopathy Journal, vol. 21, no. 2, pp. 219–226, 2008.

21. V. V. Roshchina, A. V. Yashina, V. A. Yashin, and N. K. Prizova, "Models to study pollen allelopathy,"Allelopathy Journal, vol. 23, no. 1, pp. 3–24, 2009.
22. V. V. Roshchina, A. V. Yashina, V. A. Yashin, and M. V. Gol'tyaev, "Fluorescence of biologically active compounds in plant secretory cells," in Research Methods in Plant Science, Vol. 2. Forestry and Agroforestry, S. S. Narwal, P. Pavlovic, and J. John, Eds., pp. 3–25, Studium Press, Houston, Tex, USA, 2011.
23. V. V. Roshchina and E. V. Melnikova, "Microspectrofluorometery: a new technique to study pollen allelopathy," Allelopathy Journal, vol. 3, no. 1, pp. 51–58, 1996.
24. V. V. Roshchina and E. V. Melnikova, "Chemosensory reactions at the interaction pollen-pistil," Biology Bulletin, no. 6, pp. 678–685, 1998.
25. A. Salih, A. S. Jones, D. Bass, and G. Cox, "Confocal imaging of exine as a tool for grass pollen analysis,"Grana, vol. 36, no. 4, pp. 215–224, 1997.
26. A. E. Fonseca, M. E. Westgate, and R. T. Doyle, "Application of fluorescence microscopy and image analysis for quantifying dynamics of maize pollen shed," Crop Science, vol. 42, no. 6, pp. 2201–2206, 2002.
27. V. V. Roshchina and V. N. Karnaukhov, "Changes in pollen autofluorescence induced by ozone,"Biologia Plantarum, vol. 42, no. 2, pp. 273–278, 1999.
28. V. V. Roshchina and E. V. Mel'nikova, "Pollen chemosensitivity to ozone and peroxides," Russian Journal of Plant Physiology, vol. 48, no. 1, pp. 74–83, 2001.
29. V. V. Roshchina and V. D. Roshchina, Ozone and Plant Cell, Kluwer Academic, Dordrecht, The Netherlands, 2003.
30. K. Mitsumoto, K. Yabusaki, and H. Aoyagi, "Classification of pollen species using autofluorescence image analysis," Journal of Bioscience and Bioengineering, vol. 107, no. 1, pp. 90–94, 2009.
31. K. Mitsumoto, K. Yabusaki, K. Kobayashi, and H. Aoyagi, "Development of a novel real-time pollen-sorting counter using species-specific pollen autofluorescence," Aerobiologia, vol. 26, no. 2, pp. 99–111, 2010.

32. H. Aoyagi and C. U. Ugwu, "Fulerene fine particles adhere to pollen grains and affect their autofluorescence and germination," Nanotechnology, Science and Applications, vol. 4, pp. 67–71, 2011.

33. A. Speranza, R. Crinelli, V. Scoccianti, and A. Geitmann, "Reactive oxygen species are involved in pollen tube initiation in kiwifruit," Plant Biology, vol. 14, no. 1, pp. 64–76, 2011.

34. A. A. Dobritsa, A. Geanconteri, J. Shrestha et al., "A large-scale genetic screen in Arabidopsis to identify genes involved in pollen exine production," Plant Physiology, vol. 157, no. 2, pp. 947–970, 2011.

35. A. J. Castro, J. D. Rejon, M. Fendri et al., "Taxonomical discrimination of pollen grains by using confocal laser scanning microscopy (CLSM) imaging of autofluorescence," in Microscopy: Science, Technology, Application and Education, A. Mendez-Vilas and J. Diaz, Eds., pp. 607–613, Formatex, Badajoz, Spain, 2010.

36. M. Suzuzi, M. Tonouchi, K. Muyarama, and K. Mitsumoto, "An analysis of high concentration of pollen at early morning in 2005 spring and a measurement of Japanese cedar and cypress pollen using autofluorescence characteristics," Journal of Aerosol Research, vol. 20, no. 4, pp. 281–289, 2005.

37. V. V. Roshchina, "Cellular models as biosensors," in Cell Diagnostics: Images, Biophysical and Biochemical Processes in Allelopathy, V. V. Roshchina and S. S. Narwal, Eds., pp. 5–22, Science, Enfield, UK, 2007.

38. V. V. Roshchina, "Luminescent cell analysis in allelopathy," in Cell Diagnostics: Images, Biophysical and Biochemical Processes in Allelopathy, V. V. Roshchina and S. S. Narwal, Eds., pp. 103–115, Science, Enfield, UK, 2007.

39. V. V. Roshchina, V. A. Yashin, A. V. Kononov, and A. V. Yashina, "Laser-scanning confocal microscopy (LSCM): study of plant secretory cell," in Cell Diagnostics: Images, Biophysical and Biochemical Processes in Allelopathy, V. V. Roshchina and S. S. Narwal, Eds., pp. 93–102, Science, Enfield, UK, 2007.

40. V. V. Roshchina and V. N. Karnaukhov, "The fluorescence analysis of medicinal drugs' interaction with unicellular biosensors," Farmatsiya, vol. 7, no. 3, pp. 43–46, 2010.

41. A. G. Valdecasas and A. Abad, "Morphological confocal microscopy in arthropods and the enhancement of autofluorescence after proteinase K extraction," Microscopy and Microanalysis, vol. 17, no. 1, pp. 109–113, 2011.
42. V. V. Roshchina, V. A. Yashin, and I. M. Vikhlyantsev, "Fluorescence of plant microspores as biosensors," Biochemistry Supplemental Series A, vol. 6, no. 1, pp. 105–112, 2012.
43. W. Rühland and K. Wetzel, "Der Nachweis von Chloroplasten in den generativen Zellen von Pollenschläuchen," Berirhte Der Deutarhen Botanischen Gesellschaft, vol. 42, pp. 3–14, 1924.
44. F. Berger, "Das Verhalten der heufieber-erregenden Pollen in filtrien ultravioleten Licht," Beiträge zur Biologie der Pflanzen, vol. 22, no. 1, pp. 1–12, 1934.
45. F. Asbeck, "Fluorescezieren der Blutenstaub," Naturwissenschaften, vol. 42, no. 5, p. 632, 1955.
46. P. van Gijzel, "Autofluorescence and age of some fossil pollen and spores," Proceedings of the Koninklijke Nederlandse Akademie Van Wetenschappen—Series B, vol. 64, no. 1, pp. 56–63, 1961.
47. M. T. M. Willemse, "Morphological and fluorescence microscopical investigation on sporopollenin formation at Pinus sylvestris and Gasteria verrucosa," in Sporopollenin. Proceeding Symposim at Geology Department, Imperial College, London, 1970. 23–25 September, J. Brooks, P. R. Grant, M. Muir, and P. R. van Gijzel, Eds., pp. 68–91, Academic Press, New York, NY, USA, 1971.
48. D. J. O. 'Connor, D. Iacopino, D. A. Healy, D. O'Sullivan, and J. R. Sodeau, "The intrinsic fluorescence spectra of selected pollen and fungal spores," Atmospheric Environment, vol. 45, no. 35, pp. 6451–6458, 2011.
49. Y. L. Pan, S. C. Hill, R. G. Pinnick, J. M. House, R. C. Flagan, and R. K. Chang, "Dual-excitation-wavelength fluorescence spectra and elastic scattering for differentiation of single airborne pollen and fungal particles," Atmospheric Environment, vol. 45, no. 8, pp. 1555–1563, 2011.
50. V. V. Roshchina and V. D. Roshchina, The Excretory Function of Higher Plants, Springer, Berlin, Germany, 1993.

51. V. V. Roshchina, E. V. Melnikova, V. N. Karnaukhov, and B. N. Golovkin, "Application of microspectrofluorimetry in spectral analysis of plant secretory cells," Biology Bulletin, no. 2, pp. 167–171, 1997.
52. V. V. Roshchina and E. V. Melnikova, "Spectral analysis of intact secretory cells and excretions of plants," Allelopathy Journal, vol. 2, no. 2, pp. 179–188, 1995.
53. S. Lee, R. L. Brown, and W. Monroe, "Use of confocal laser scanning microscopy in systematics of insects with a comparison of fluorescence from different stains," Systematic Entomology, vol. 34, no. 1, pp. 10–14, 2009.
54. Z. Žižka and J. Gabriel, "Primary fluorescence (autofluorescence) of fruiting bodies of the wood-rotting fungus Fomes fomentarius," Folia Microbiologica, vol. 51, no. 2, pp. 109–113, 2006.
55. C. H. Wu and H. L. Warren, "Natural autofluorescence in fungi and its correlation with viability,"Mycologia, vol. 76, no. 6, pp. 1049–1058, 1984.
56. C. H. Wu and H. L. Warren, "Induced autofluorescence in fungi, and its correlation with viability: potential application of fluorescence microscopy," Phytopathlogy, vol. 74, no. 6, pp. 1353–1358, 1984.
57. G. Doehlemann, K. van der Linde, D. Aßmann et al., "Pep1, a secreted effector protein of Ustilago maydis, is required for successful invasion of plant cells," PLoS Pathogens, vol. 5, no. 2, Article ID e1000290, 2009.
58. A. R. Hardham, "Confocal microscopy in plant-pathogen interactions," in Plant Fungal Pathogens: Methods and Protocols, M. D. D. Bolton, P. H. J. Bart, and P. Thomma, Eds., vol. 835 of Methods in Molecular Biology, pp. 295–309, Springer, Berlin, Germany, 2012.
59. F. Perrine-Walker, P. Doumas, M. Lucas et al., "Auxin carriers localization drives auxin accumulation in plant cells infected by Frankia in Casuarina glauca actinorhizal nodules," Plant Physiology, vol. 154, no. 3, pp. 1372–1380, 2010.
60. F. Gandía-Herrero, F. García-Carmona, and J. Escribano, "Botany: floral fluorescence effect," Nature, vol. 437, no. 7057, p. 334, 2005.

61. P. Hutzler, R. Fischbach, W. Heller et al., "Tissue localization of phenolic compounds in plants by confocal laser scanning microscopy," Journal of Experimental Botany, vol. 49, no. 323, pp. 953–965, 1998.
62. G. Agati, Z. G. Cerovic, P. Pinelli, and M. Tattini, "Light-induced accumulation of ortho-dihydroxylated flavonoids as non-destructively monitored by chlorophyll fluorescence excitation techniques,"Environmental and Experimental Botany, vol. 73, pp. 3–9, 2011.
63. G. Conejero, P. Talamond, and J. L. Verdeil, "A new approach to visualize secondary metabolites in plants. Histocytology and plant cell imaging control platform," La plate-forme regionale d›imagerie "Montpellier Rio Imaging" (MRI) organise une journee "Imagerie", Lafayette, La, USA, Octobre 2009,http://phiv.cirad.fr/download/poster.pdf.
64. A. Iriel and M. G. Lagorio, "Implications of reflectance and fluorescence of Rhododendron indicumflowers in biosignaling," Photochemical and Photobiological Sciences, vol. 9, no. 3, pp. 342–348, 2010.
65. A. Iriel and M. G. Lagorio, "Is the flower fluorescence relevant in biocommunication?"Naturwissenschaften, vol. 97, no. 10, pp. 915–924, 2010.
66. F. Gandía-Herrero, J. Escribano, and F. García-Carmona, "Betaxanthins as pigments responsible for visible fluorescence in flowers," Planta, vol. 222, no. 4, pp. 586–593, 2005.
67. F. Gandía-Herrero, M. Jiménez-Atiénzar, J. Cabanes, J. Escribano, and F. García-Carmona, "Fluorescence detection of tyrosinase activity on dopamine-betaxanthin purified from Portulaca oleracea (common purslane) flowers," Journal of Agricultural and Food Chemistry, vol. 57, no. 6, pp. 2523–2528, 2009.
68. G. Agati, M. L. Traversi, and Z. G. Cerovic, "Chlorophyll fluorescence imaging for the noninvasive assessment of anthocyanins in whole grape (Vitis Vinifera L.) bunches," Photochemistry and Photobiology, vol. 84, no. 6, pp. 1431–1434, 2008.
69. D. L. Betemps, J. C. Fachinello, S. P. Galarca et al., "Non-destructive evaluation of ripening and quality traits in apples using a multiparametric fluorescence sensor," Journal of the

Science of Food and Agriculture, vol. 92, no. 9, pp. 1855–1864, 2012.

70. C. Gomez, G. Conejero, L. Torregrosa, V. Cheynier, N. Terrier, and A. Ageorges, "In vivo grapevine anthocyanin transport involves vesicle-mediated trafficking and the contribution of anthoMATE transporters and GST," Plant Journal, vol. 67, no. 6, pp. 960–970, 2011.
71. V. V. Roshchina, "Plant microspores as biosensors," Trends in Modern Biology, vol. 126, no. 3, pp. 262–274, 2006.
72. V. V. Roshchina, E. V. Mel›nikova, V. A. Yashin, and V. N. Karnaukhov, "Autofluorescence of intactEquisetum arvense L. spores during their development," Biofizika, vol. 47, no. 2, pp. 318–324, 2002.
73. V.V. Roshchina, V. A. Yashin, and A. V. Kononov, "Autofluorescence of developing plant vegetative microspores studied by confocal microscopy and microspectrofluorimetry," Journal of Fluorescence, vol. 14, no. 6, pp. 745–750, 2004.
74. V. V. Roshchina, Neurotransmitters in Plant Life, Science, Enfield, UK, 2001.
75. V.V. Roshchina, "Evolutionary considerations of neurotransmitters in microbial, plant and animal cells," in Microbial Endocrinology: Interkingdom Signaling in Infectious Disease and Health, M. Lyte and P. P. E. Freestone, Eds., pp. 17–52, Springer, Berlin, Germany, 2010.
76. R. N. Ames, E. R. Ingham, and C. P. P. Reid, "Ultraviolet-induced autofluorescence of a mycorrhizal root infestions and alternative to clearing and staining methods for infections," Canadian Journal of Microbiology, vol. 28, pp. 351–355, 1982.
77. B. Dreyer, A. Morte, M. Pérez-Gilabert, and M. Honrubia, "Autofluorescence detection of arbuscular mycorrhizal fungal structures in palm roots: an underestimated experimental method," Mycological Research, vol. 110, no. 8, pp. 887–897, 2006.
78. B. Dreyer, "Use of the autofluorescence properties of AM fungi for AM assessment and handling," Soil Biology, vol. 18, pp. 123–140, 2009.

Chapter 2

# Imaging of Chlorophyll a Fluorescence: A Tool to Study Abiotic Stress in Plants

Lucia Guidi[1] and Elena Degl'Innocenti[1]

[1]Dipartimento di Biologia delle Piante Agrarie, Università di Pisa, Italy

## INTRODUCTION

Chlorophyll (Chl) fluorescence is a tool which is widely used to examine photosynthetic performance in algae and plants. It is a non-invasive analysis that permits to assess photosynthetic performance in vivo (Baker, 2008, Baker & Rosenqvist, 2004, Chaerle & Van Der Straeten, 2001, Woo et al. 2008). Chl fluorescence analysis is widely used to estimate photosystem II (PSII) activity, which is an important target of abiotic stresses (Balachandran et al., 1994, Baker et al., 1983,

Briantais et al., 1996, Calatayud et al., 2008, Chaerle & Van Der Straeten, 2000, Ehlert & Hincha, 2008; Gilmore & Govindjee, 1999, Guidi et al., 2007 , Guidi & Degl'Innocenti, 2008, Hogewoning & Harbinson, 2007,Krause, 1988, Lichtenthaler et al., 2007; Massacci et al., 2008, Osmond et al., 1999, Scholes & Rolfe, 1996, Strand & Oquist, 1985).

It is known as the energy absorbed by Chl molecules must be dissipated into three mechanisms, namely internal conversion, fluorescence and photochemistry (Butler, 1978). All of these downward processes competitively contribute to the decay of the Chl excited state and, consequently, an increase in the rate of one of these processes would increase its share of the decay process and lower the fluorescence yield. Typically, all processes that lower the Chl fluorescence yield are defined with the term*quenching*.

Kaustky and co-workers (1960) were the first which observed changes in yield of Chl fluorescence. These researchers found that transferring a leaves from the dark into the light, an increase in Chl fluorescence yield occurred. This increase has been explained with the reduction of electron acceptors of the PSII and, in particular, plastoquinone $Q_A$: once PSII light harvesting system (LCHII) absorbs light and the charge separation occurs, $Q_A$ accepts electron and it is not able to accept another electron until it has been passed the first one onto the subsequent carrier, namely plastoquinone $Q_B$. During this time the reaction centers are said to be *closed*. The presence of closed reaction centers determines a reduction in the efficiency of PSII photochemistry and, consequently, an increase in the Chl fluorescence yield.

Transferring the leaf from the dark into light, PSII reaction centers are progressively closed, but, following this time, Chl fluorescence level typically decreases again and this phenomenon is due to two types of quenching mechanisms. The presence of light induced the activation of enzymes involved in $CO_2$ assimilation and the stomatal aperture that determines that electrons are transferred away PSII. This induced the so-called *photochemical quenching*, $q_P$. At the same time, there is an increase in the conversion of light energy into heat related to the *non-photochemical quenching*, $q_{NP}$. This non-photochemical quenching $q_{NP}$, can be divided into three components. The major and most rapid component in algae and plants is the pH- or energy-dependent component, $q_E$. A second component, $q_T$, relaxes within minutes and is due to the phenomenon of state transition, the uncoupling of LHCIIs from PSII. The third component of $q_{NP}$ shows the slowest relaxation and is the least defined. It is related to photoinhibition of photosynthesis and is therefore called $q_I$.

To evaluate Chl fluorescence quenching coefficients during illumination we must determine minimal and maximal fluorescence yields after dark adaptation, $F_0$ and $F_m$ respectively. This is important because these values serve as references for the evaluation of the photochemical and non-photochemical quenching coefficients in an illuminated leaf by using the *saturation pulse method*. The concept on the basis of this method is extremely simply: at any give state of illumination, $Q_A$ can be fully reduced by a saturation pulse of light, such that photochemical quenching is completely suppressed. During the saturation pulse, a maximal fluorescence $F_m'$ is achieved which generally shows value lower that the dark reference values ($F_m$). With the assumption that non-photochemical quenching does not change during a short saturation pulse, the reduction of $F_m$ is a measure of non-photochemical quenching.

In Figure 1 the calculation of Chl fluorescence parameters by using the saturation pulse method is reported. The photochemical quenching coefficient $q_P$ is measured as

$$q_P = (F_m' - F_t) / (F_m' - F_0') \tag{1}$$

where $F_m'$ is the maximum Chl fluorescence yield in light conditions, $F_t$ is the steady-state Chl fluorescence immediately prior to the flash. For determination of $F_0'$ in the light state, the leaf has to be transiently darkened and it has to be assured that $Q_A$ is quickly and fully oxidized, before there is a substantial relaxation of non-photochemical quenching. In order to enhance of oxidation of the intersystem electron transport chain, far-red light is applied that selectively excited PSI. Usually the alternative expression of this quenching coefficient is used and it is (1- $q_p$). i.e. the proportion of centers that are closed and it is termed *excitation pressure* on PSII (Maxwell & Johnson, 2000).

Another useful fluorescence parameter derived from saturation pulse method is the efficiency of PSII photochemistry, which is calculated as:

$$F_{PSII} = (F_m' - F_t) / F_m' \qquad (2)$$

This parameter has also termed $F/F_m'$ or, in fluorescence imaging technique, $F_q'/F_m'$ and it is very similar to the $q_P$ coefficient even if its significance is somewhat different. The $_{PSII}$ is the proportion of absorbed light energy being used in photochemistry, whilst $q_P$ gives an indication of the proportion of the PSII reaction centers that are open. A parameter strictly related with both $q_P$ and $_{PSII}$ is the ratio $F_V/F_m$ determined as:

$$F_v / F_m = (F_m - F_0) / F_m \qquad (3)$$

This third parameter is determined in dark adapted leaves and it is a measure of the maximum efficiency of PSII when all centers are open. This ratio is a sensitive indicator of plant photosynthetic performance because of it has an optimal values of about 0.83 in leaves of healthy plants of most species (Bjorkman & Demmig, 1987). Another useful parameter which describes energy dissipation is $F_V'/F_m'$, an estimate of the PSII quantum efficiency if all PSII reaction centers are in the open state. It is calculated as reported in equation 4:

$$F_v' / F_m' = (F_m' - F_0') / F_m' \qquad (4)$$

Since $_{PSII}$ is the quantum yield of PSII photochemistry, it can be used to determine linear electron transport rate (ETR) as described by Genty et al., (1989):

$$ETR = \phi_{PSII} x PPFD x 0.5 \qquad (5)$$

where PPFD (photosynthetic photon flux density) is the absorbed light and 0.5 is a factor that accounts for the partitioning of energy between PSII and PSI.

The excess of excitation energy which is not used for photochemistry can be de-excited by thermal dissipation processes.

Non-photochemical quenching of Chl fluorescence is an important parameter that gives indication of the non-radiative energy dissipation in the light-harvesting antenna of PSII. This parameter is extremely important taking into account that the level of excitation energy in the antenna can be regulated to prevent over-reduction of the electron transfer chain and protect PSII from photodamage. Non-photochemical quenching coefficient is calculated as:

$$q_{NP} = (F_m - F_m') / (F_m - F_0') \quad (6)$$

In some circumstances $F_0'$ determination is difficult, e.g. in the field when a leaf cannot be transiently darkened. In this case, another parameter can be used to describe non-photochemical energy dissipation NPQ (Schreiber & Bilger, 1993), which does not require the knowledge of $F_0'$. The parameter NPQ is derived from Stern-Volmer equation and its determination implies the assumption of the existence of traps for nonradiative energy dissipation, like zeaxanthin, in the antenna pigment matrix (Butler, 1978). NPQ is calculated as reported in equation 7 (Bilger & Bjorkman, 1990):

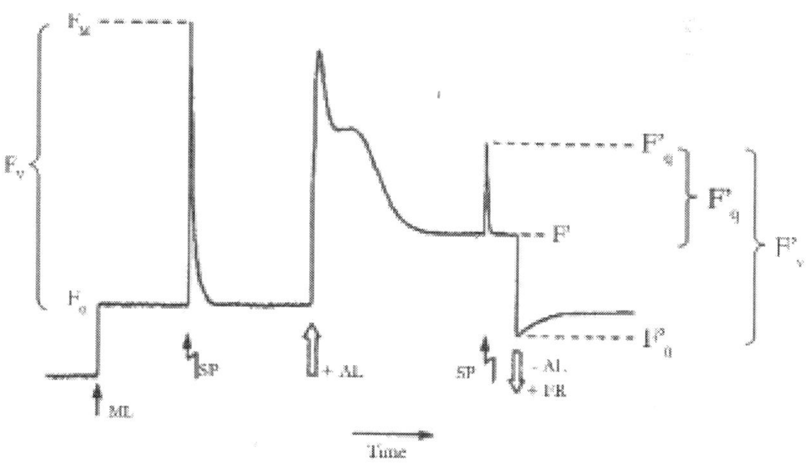


**Figure 1:** Measurement of chlorophyll fluorescence by the saturation pulse method (adapted from Van Kooten & Snell, 1990).

$$NPQ = (F_m - F_m') / F_m' \quad (7)$$

NPQ is linearly related to heat dissipation and varies on a scale from 0 until infinity even if in a typical plants value ranges between 0.5 and 3.5 at light saturation level.

Chl fluorescence analysis gives a measure of the photosynthetic rate and for this reason it is extremely useful. Really, Chl fluorescence gives information about the efficiency of PSII photochemistry that, in laboratory conditions, is strictly correlated with $CO_2$ photoassimilation (Edwards & Baker, 1993,Genty et al., 1989). Under field conditions, this correlation is lost because other processes compete with $CO_2$ assimilation such as photorespiration, nitrogen metabolism and Mehler reaction (Fryer et al., 1998). In addition to, a complication derives to heterogeneity between samples. To calculate ETR we assume that the light absorbs by PSII is constant, but it is not true. Even if there are some limitations, Chl fluorescence can give a good, rapid and non-invasive measurements of changes in PSII photochemistry and then also the possibility to evaluate the effects of abiotic stresses on PSII performance.

# CHL FLUORESCENCE IMAGING

The evolution of Chl fluorescence analysis is represented by Chl fluorescence imaging which can be useful applied into two general areas: the study of heterogeneity on leaf lamina and the screening of a large numbers of samples. This technique has been widely applied in the past during induction of photosynthesis (Bro et al., 1996, Oxborough & Baker, 1997), with changes in carbohydrate translocation (Meng et al., 2001), in response to drought (Meyer & Genty, 1999, West et al., 2005), chilling (Hogewoning & Harbinson, 2007), ozone pollution (Guidi et al., 2007, Guidi & Degl'Innocenti, 2008, Leipner et al., 2001), wounding (Quilliam et al., 2006), high light (Zuluaga et al., 2008) and infection with fungi (Guidi et al., 2007, Meyer et al., 2001, Scharte et al., 2005,Scholes & Rolfe, 1996, Schwarbrick et al., 2006) or virus (Perez-Bueno et al., 2006). With Chl fluorescence imaging is possible to detect an analysis of stress-induced changes in fluorescence emission at very early stage of stress. In addition to, Chl fluorescence imaging technique represents a useful screening tool for crop yield improvement.

The most essential new information provided by Chl fluorescence imaging relates to the detection of lateral heterogeneities of fluorescence parameters which reflect physiological heterogeneities. It is well known that even physiologically healthy leaves are "patchy" with respect to stomatal opening. Furthermore, stress induced limitations, which eventually will lead to damage, are not evenly distributed over the whole leaf area. Fluorescence imaging may serve as a convenient tool for early detection of such stress induced damage. The main difference between the conventional fluorometer and the imaging fluorometer is the possibility of parallel assessment of several samples under identical conditions.

For example we treated plants of *Phaseolus vulgaris* (cv. Cannellino) with a single pulse of ozone ($O_3$) (150 nL $L^{-1}$for 5 h) and evidenced upon leaf lamina and evident heterogeneity in some Chl fluorescence parameter as compared to control exposed to charcoal filtered air for the same period (Guidi & Degl'Innocenti, *data not published*) (Figure 2).

It is know as in plants exposed to chilling stress, photosynthetic enzymes may be inactivated or degraded and photodamage to PSII may happen, reducing photosynthesis (Dai et al., 2007, Feng & Cao, 2005, Flexas et al. 1999). The reduction in photosynthetic $CO_2$ assimilation may lead to accumulation of excess energy especially at high irradiance and consequently to photoinhibition (Feng & Cao, 2005, Hovenden & Warren, 1998). In variegated leaves of *Calathea makoyana* the effect of chilling (5 and 10 C for 1-7 d) on PSII efficiency was studied in order to understand the causes of chilling-induced photoinhibition (Hogewoning & Harbinson, 2007). The individual leaves were divided into a shaded zone and two illuminated, chilled zones. Chilling up to 7 d in the dark did not influence PSII efficiency whereas chilling in the light caused severe photoinhibition. Data obtained from Chl fluorescence imaging were confirmed by visual appearance of symptoms which were evident in the portion of leaves chilled and illuminated. Obtained results showed that photoinhibition was due to a secondary effect in the unchilled leaf tip (sink limitation) as revealed by starch accumulation data. Instead it was a direct effect of chilling and irradiance in the chilled illuminated zones.

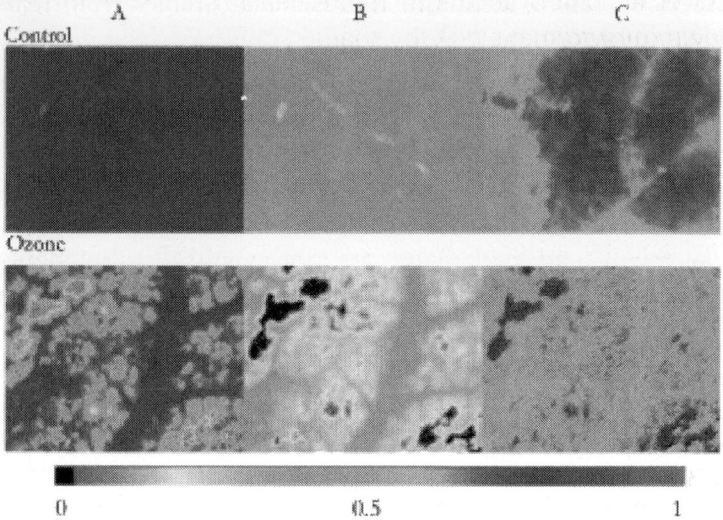


**Figure 2:** Chl fluorescence imaging of $F_v/F_m$ (A), $_{PSII}$ (B) and non-photochemical quenching (C) in leaves of*P. vulgaris* cv. Cannellino exposed for 5 h at an $O_3$ concentration of 150 nL $L^{-1}$ (Ozone) or 2 nL $l^{-1}$(Control). All images are normalised to the false colour bar provided. The analyses of $F_v/F_m$ were carried out on dark-adapted leaves, while $_{PSII}$ and $q_{NP}$ at a light intensity of 500 mol $m^{-2}s^{-1}$. The pixel value display is based on a false-colour scale ranging from black (0.00 to 0.040) via red, yellow, green, blue to purple (ending at 1.00) (from Guidi & Degl'Innocenti, *data not published*).

Calatayud et al. (2008) studied the effects of two nutrient solution temperatures (10 and 22 C) during the flowering of *Rosa x hybrida* by using Chl fluorescence imaging. The obtained results showed as the nutrient solution temperatures of 10 C induced an increase in $_{PSII}$ parameters indicating that the majority of photons absorbed by PSII were used in photochemistry and that PSII centers were maintained in an oxidized state.

Water stress is another important abiotic stress that induces reduction of growth and yield of plants. For this reason the development of drought-tolerance is an important target of the researchers. The effects of drought on photosynthetic process have been extensively studied in many plant species and the possible mechanisms involved in the responses have been suggested (Cornic & Fresneau, 2002,Flexas et al., 2002, 2004, Grassi & Magnani, 2005, Long & Bernacchi, 2003).

Masacci et al. (2008) took Chl fluorescence images from leaves of *Gossypium hirsutum* to study the spatial pattern of PSII efficiency and non-photochemical quenching parameters. They found that under low and moderate light intensity, the onset of drought stress caused an increase in the operating quantum efficiency of PSII ($_{PSII}$) which indicated increased photorespiration since photosynthesis was hardly affected by water shortage. The increase in $_{PSII}$ was caused by an increase in $F_v'/F_m'$ and by a decrease in non-photochemical quenching. Chl fluorescence imaging showed a low spatial heterogeneity of $_{PSII}$. The authors concluded that the increase in photorespiration rate in plants during the water stress can be seen as an acclimation process to avoid an over-excitation of PSII under more severe drought conditions.

Qing-Ming et al., (2008) used Chl fluorescence imaging analysis to detect the effects of drought stress and elevated $CO_2$ concentration (780 mol mol$^{-1}$) in cucumber seedlings. They found that electron transport rate and the light saturation level declined significantly with drought stress aggravation in both $CO_2$ concentrations. Drought stress decreased maximal photosynthetic ETR and subsequently decreased the capacity of preventing photodamage. At the same time, elevated $CO_2$ concentration increased the light saturation level significantly, irrespective of the water conditions. Elevated $CO_2$concentration can alleviate drought stress-induced photoinhibitory damage by improving saturating photosynthetically active radiation.

Sommerville et al. (2010) examined the different spatial response in photosynthesis with drought in two species with contrasting hydraulic architecture. The authors hypothesized that areole regions near primary nerves would show a smaller decline in the maximum efficiency of PSII photochemistry with drought compared with regions between secondary nerves and that the difference between areole regions would be smaller in phyllodes with higher primary nerve density. Indeed, the phyllodes of*Acacia floribunda* were found to have both greater primary nerve density and show greater spatial homogeneity in photosynthetic function with drought compared with the phyllodes of *Acacia pycnantha*. *A. floribunda* phyllodes also maintained function of the photosynthetic apparatus with drought for longer and recovered more swiftly from drought than *A. pycnantha*.

Drought is a type of stress which can induce heterogeneity in leaf photosynthesis that probably occurs when dehydration is rapid as in the case of drought experiments performed on potted plants by withholding water. Using Chl fluorescence imaging, Flexas et al. (2006) showed in herbaceous species that exogenous ABA did not induce patchy stomatal closure even when stomatal conductance dropped too much lower values lower than 0.05 mol $m^{-2}$ $s^{-1}$.

Even the quality and quantity of light intensity notable influence the photosynthetic apparatus and functioning. Generally, sun- and shade leaves differ in the composition of leaf pigment, electron carriers on thylakoids membranes, structure of the chloroplast and photosynthetic rate (Anderson et al., 1995, Boardman, 1977, Lichtenthaler, 1981, 1984, Lichtenthaler et al., 2007, Takahashi & Badger, 2010). Lichtenthaler et al. (2007) studied the differential pigment composition and photosynthetic activity of sun and shade leaves of deciduous (*Acer psuedoplatanus, Fagus sylvatica, Tilia cordata*) and coniferous (*Abies alba*) trees by using Chl fluorescence imaging analysis. This tool not only provided the possibility to screen the differences in photosynthetic $CO_2$ assimilation rate between sun and shade leaves, but in addition permitted detection and quantification of the large gradient in photosynthetic rate across the leaf area existing in sun and shade leaves.

Chl fluorescence analysis is used also to characterized photosynthetic process in transgenic plants such as tomato (*Lycopersicon esculentum*) cv. Micro-Tom transformed with the *Arabidopsis thalianaMYB75/PAP1* (PRODUCTION OD ANTHOCYANIN PIGMENT 1) gene (Zuluaga et al., 2008). This gene encodes for a well known transcription factor, which is involved in anthocyanin production and is modulated by light and sucrose. The presence of a higher constitutive level of anthocyanin pigments in transgenic plants could give them some advantage, in terms of adaptation and defence against environmental stresses. To test this hypothesis, a high light experiment was carried out exposing wild type and transgenic tomato plants to a strong light irradiance for about ten days and monitoring the respective phenotypic and physiological changes. The light intensity used was very high and likely not similar to normal environmental conditions (at least for such a prolonged period). Chlorophyll fluorescence imaging on control and stressed leaves from both genotypes suggest that, in transgenic leaves, the apparent tolerance to photoinhibition was probably not due to an increased capacity for PSII to repair, but reflected instead the ability of these leaves to protect their photosynthetic apparatus.

Certainly among abiotic stress the pollutants can alter the physiology and biochemistry of plants. Ozone is an air pollutant that induces reduction in growth and yield of plants species. The major target of the $O_3$ effects is represented by photosynthetic process and many works have been reported as this pollutant can impair $CO_2$ assimilation rate. Plant response depends also on the dose (concentration x time). In fact, it can distinguish chronic exposure to $O_3$ from acute one. It is termed chronic exposure the long-term exposure at concentration < 100 nL $L^{-1}$ whereas the acute $O_3$ exposure is generally defined as exposure to a high level of $O_3$ concentration (> 100 nL $L^{-1}$) for a short period of time, typically on the order of hours (Kangasjarvi et al., 2005). Chen et al. (2009) studied the effects of acute (400 nL $L^{-1}$, 6 h) and chronic (90 nL $L^{-1}$, 8 h $d^{-1}$, 28 d) $O_3$ concentration on photosynthetic process of soybean plants. Although both acute and chronic $O_3$ treatment resulted in a similar overall photosynthetic impairment compared to the controls, the fluorescence imaging analysis revealed that the physiological mechanisms underlying the decreases differed. In the acute $O_3$ treatments over the chronic one there was a greater spatial heterogeneity related to several bases. The higher $O_3$concentration typically induced oxidative stress and the hypersensitive response within a matter of hours leading to programmed cellular death (PCD). By the end of chronic $O_3$ treatment, control leaves showed an increase in spatial heterogeneity of photosynthesis linked to the process of natural senescence. Clearly, in this study it has been demonstrated as Chl fluorescence imaging represents a useful tool to study also mechanisms on the basis of plants responses to abiotic stress such as $O_3$pollution.

Guidi et al. (2007) used Chl fluorescence analysis to study the effects of an acute $O_3$ treatment (150 nL $L^{-1}$ for 5 h) or artificial inoculation with a pathogen (*Pleiochaeta setosa*) on photosynthesis of *Lupinus albus*. The aim of the work was to compare the perturbations in photosynthesis induced by an abiotic or biotic stress. In addition to, in the work were compared results obtained by conventional Chl fluorescence analysis and the technique of Chl fluorescence imaging. Image analysis of $F_v/F_m$ showed a different response in plants subjected to ozone or inoculated with *P. setosa*. Indeed, in ozonated leaves fluorescence yield was lower in leaf veins than in the mesophyll with the exception of the necrotic areas where no fluorescence signals could be detected. This suggests that the leaf area close to the veins were more sensitive to ozone. The parameter $_{PSII}$ decreased significantly in both infected and ozonated leaves, but image analysis provides more information than the conventional fluorometer. In fact, until 48 h after ozone treatment or fungal inoculation, $_{PSII}$ tended to decrease, especially in the infected leaves. Afterwards, a distinct stimulation of photosynthesis was observed in the area surrounding the visible lesions induced by the fungus. This did not occur in the ozonated leaves, as suggested also by the higher values of qP (data not shown). This phenomenon was not observed using the conventional fluorometer which recorded a similar reduction in this parameter in both ozonated and inoculated leaves.

In an other work Guidi and Degl'Innocenti (2008) studied the response to photoinhibiton and subsequent recovery in plants of *Phaseolus vulgaris* (cv. Pinto) exposed to charcoal-filtered air or to an acute $O_3$ exposure (150 nL $L^{-1}$ for 3 or 5 h). Susceptibility to photoinhibition in bean leaves was determined as changes in the $F_v/F_m$ ratio and the images of the ratio are reported in Figure 3. Initial values of $F_v/F_m$ were 0.796, 0.784 and 0.741 for plants maintained in charcoal-filtered air, or treated with a single exposure to $O_3$ for 3 h, or for 5 h, respectively. The results indicate that treatment with $O_3$ for 5 h induced a slight photoinhibition. The exposure of control plants (charcoal-filtered air for 5 h) at a light intensity of 1000 mol $m^{-2}s^{-1}$ resulted in a significant reduction in $F_v/F_m$ ($P < 0.01$) (Fig. 2b), while plants treated with $O_3$ for 3 h showed an increased tolerance to photoinhibition with less reduction in $F_v/F_m$ (Fig. 2f). Plants treated with $O_3$ for 5 h and then exposed to high light showed a reduction in $F_v/F_m$ ratio values similar to those recorded in control plants (Fig. 2i and l). However, while control plants or treated with $O_3$ for 3 h recovered their initial value 24 h after photoinhibition treatment, plants treated with $O_3$ for 5 h did not show the same ability to recover. In these plants the values of the $F_v/F_m$ ratio did not recover and, 48 h after photoinhibition leaves showed visible symptoms of damage over the entire surfaces which precluded further analysis. At the same time, severe wilting did not permit chlorophyll fluorescence imaging.

Most of the abiotic stresses induce in plants an oxidative damage of the cell structure and consequently a loss in the cellular activities. Chloroplast represents the organelle which possesses pigments that absorb light and drive redox reactions of thylakoids but also the site in the cell where $O_2$ is evolved from water. Clearly, it represents an organelle such as mitochondria, in which the formation of reactive oxygen species (ROS) can occur. On the other hand, chloroplasts are able to produce strong oxidants associated with PSII which are responsible for the splitting of $H_2O$ molecules, but they can also oxidize pigments, proteins and lipid of the thylakoid membranes as well. This characteristic makes the chloroplast a major stress sensor in green plants (Biswal & Biswal 1999). Even the separation charge and the electron transport rate associated represent another important factor that makes chloroplast sensitive to stress. Using image analysis tools Aldea et al. (2006) observed a statistical relationship between ROS and reductions in photosynthetic efficiency ($_{PSII}$) in leaves damaged simultaneously by $O_3$ (80 nL $L^{-1}$ for 8 h) and viral infection (soybean mosaic virus). The author by using Chl fluorescence analysis overlapped spatial maps of $_{PSII}$ and ROS and found that areas with depressed $_{PSII}$ corresponded to areas of high ROS concentration.

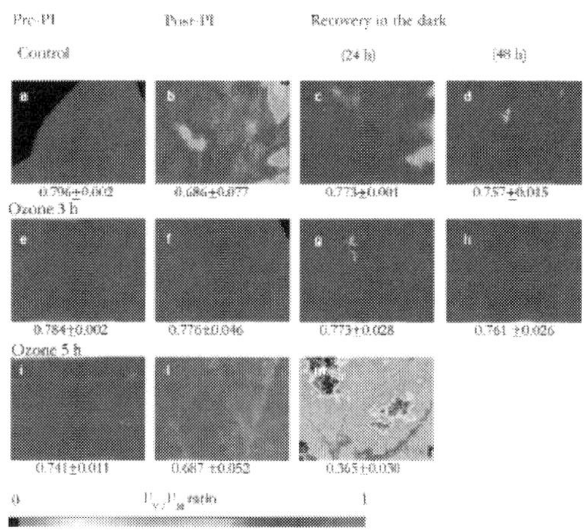


**Figure 3:** Representative fluorescence images of the Fv/Fm ratio in leaves of *Phaseolus vulgaris* L. cultivar Pinto after a single exposure to $O_3$ (150 ppb) for 3 h (Ozone 3 h; e–h) or 5 h (Ozone 5 h; i–m) or exposed to charcoal-filtered air (control, a–d) (Pre-PI). The images correspond to different measurement times: after charcoal-filtered air or $O_3$ exposure (a, e and i), after photoinhibitory treatment for 5 h (b, f and l), after recovery in the dark for 24 h (c, g and m) or for 48 h (d and h). All images are normalised to the false colour bar provided. The analyses of Fv/Fm were carried out on dark-adapted leaves. The pixel value display is based on a false-colour scale ranging from black (0.00 to 0.040) via red, yellow, green, blue to purple (ending at 1.00) (from Guidi & Degl'Innocenti, 2008).

Wounding is another common abiotic stress which induces a spatial and temporal complex series of responses in plants. In fact, wounding induces by herbivore or mechanical damage determines localized cell death, loss of water and solutes from cut surface which provides a point of entry of bacterial and fungal pathogens and disrupts vascular system. Many responses can be activated following wounding such as defense and repair mechanisms which require a high metabolic demand upon wounded region. These responses determine the synthesis of new molecules and then energy and carbon skeleton. An interesting work reported the study of the spatial and temporal changes in source-sink relationships which occur in mechanically wounded leaves of *Arabidopsis thaliana* (Quilliam et al., 2006). When the Chl fluorescence imaging analyses was made immediately after wounding there was a localized reduction in the steady-state of $_{PSII}$ in cells adjacent to the wound margin and this suggests that these cells were damaged. No changes in $F_v/F_m$ ratio were observed. Twenty-four hours after wounding, cells proximal to the wound margin showed a rapid induction of $_{PSII}$ upon illumination whilst cells more distal to the wound margin exhibited a much slower induction of $_{PSII}$ and a large increase of NPQ. The obtained results indicate of an increase in sink strength in the vicinity of the wound.

Chl fluorescence imaging has been used also for particular studies such as the characterization of a mutants with altered leaf morphology that are useful as markers for the study of genetic systems and for probing the leaf differentiation process. In a study carried out by Fambrini et al (2010) a mutant with deficient greening and altered development of the leaf mesophyll appeared in an inbred line of sunflower (*Helianthus annuus* L.). The mutation, named *mesophyll cell defective1* (*mcd1*), has pleiotropic effects and it is inherited as a monogenic recessive. The structure and tissue organization of*mcd1* leaves are disrupted A deficient accumulation of photosynthetic pigments characterizes both cotyledons and leaves of the mutant. In *mcd1* leaves, Chl fluorescence imaging evidences a spatial heterogeneity of leaf photosynthetic performance. Little black points, which correspond to PSII maximum efficiency ($F_v/F_m$) values close to zero, characterize the *mcd1* leaves. Similarly, the light adapted quantum efficiency ($\Phi_{PSII}$) values show a homogeneous distribution over wild type leaf lamina, while the damaged areas in *mcd1* leaves, represented by yellow zones, are prominent (Figure 4).

In conclusion, the loss of function of the *MCD1* gene in *Helianthus annuus* is correlated with a variegated leaf phenotype characterized by a localized destruction of mesophyll morphogenesis and defeat of PSII activity.

Another interesting application of Chl fluorescence imaging in represented by its used to analyze the generation of action potentials in irritated *Dionaea muscipula* traps to determine the 'site effect' of the electrical signal-induced inhibition of photosynthesis (Pavlovic et al. 2011). Irritation of trigger hairs and subsequent generation of action potentials resulted in a decrease in the effective photochemical quantum yield of photosystem II ($_{PSII}$) and the rate of net photosynthesis (Figure 5).

During the first seconds of irritation, increased excitation pressure in PSII was the major contributor to the decreased $_{PSII}$. Within 1 min, NPQ released the excitation pressure at PSII. All the data presented in this work indicate that the main primary target of the electrical signal induced inhibition of photosynthesis is the dark reaction, whereas the inhibition of electron transport is only a consequence of reduced carboxylation efficiency. In addition, the study also provides valuable data confirming the hypothesis that chlorophyll a fluorescence is under electrochemical control.

Chl fluorescence imaging combined with thermal imaging has been used also for monitoring and screening plant population (Chaerle et al., 2006). Rapid screening for stomatal responses can be achieved by thermal imaging, while, combined with fluorescence imaging to study photosynthesis, can potentially be used to derive leaf water use efficiency as a screening parameter.

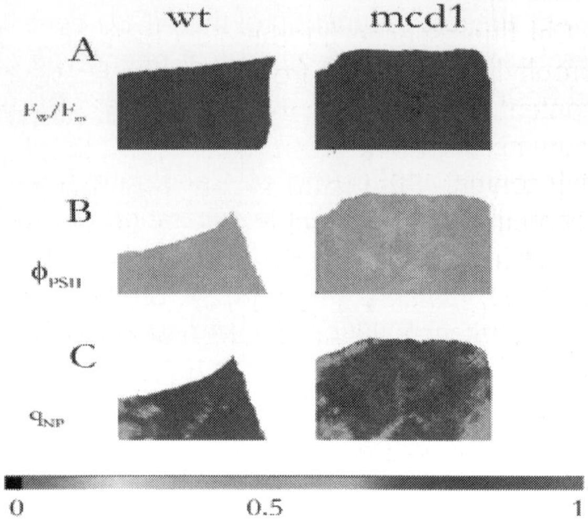


**Figure 4:** Analysis of chlorophyll fluorescence parameters in wild type (wt) and *mesophyll cell defective1(mcd1)* mutant plants of sunflower (*Helianthus annuus* L.). *A–C*: Fluorescence images of the maximum efficiency of PSII ($F_v/F_m$; *A*), the proportion of absorbed light, which is utilized for photosynthetic electron transport (ΦPSII; *B*), and the nonphotochemical quenching coefficient (qNP;*C*), in representative leaves from wild type (*left column*) and *mcd1* mutant (*right column*), are shown (from Fambrini et al., 2010).

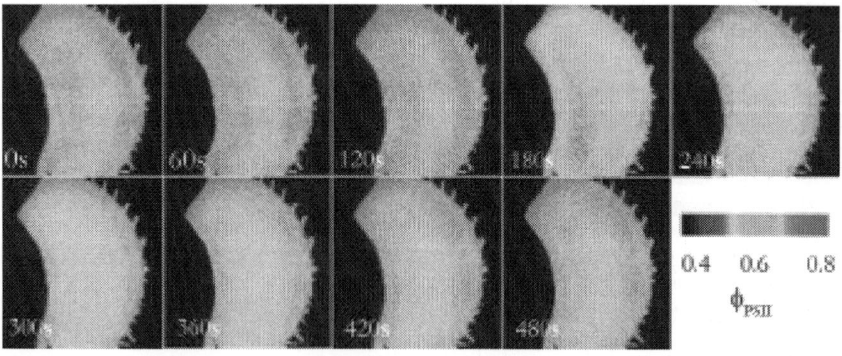


**Figure 5:** Spatiotemporal changes of effective photochemical quantum yield of PSII ($\Phi_{PSII}$) in a *D. muscipula*closed trap assessed by chlorophyll fluorescence imaging. The trap was irritated by a thin wire between 162 s and 177 s (image obtained from Pavlovic et al. 2011).

Although Chl fluorescence fluorometers have been developed to measure chlorophyll fluorescence from green tissues, which are high in chlorophyll content, the extraordinary sensitivity of current instruments enables measurements in non-green plant tissues that have relatively low chlorophyll content. This includes many types of ripening fruit that during development degrade the chloroplasts (including chlorophyll) that are contained in the fruit skin. Even non-green fruit that are highly colored (e.g., apples, tomatoes), contain active chloroplasts that yield a chlorophyll fluorescence signal of sufficient strength that it can be used as a probe of photosynthetic activity in the fruit skin (DeEll et al., 1995). In food technology, Chl fluorescence imaging can provide a rapid and non-invasive, post-harvest evaluation of the quality of fruits and vegetables (DeEll et al., 1995, DeEll & Toivonen, 2000). Nedbal & Withmarsh (2004) reported an interesting article on this topic. By applying fluorescence imaging on individual fruit before any symptoms of bitter pit were apparent, lower fluorescence was shown to be associated with bitter pit development in apples in selective cases (Lotze et al., 2006). The authors showed that, using averaged cumulative distribution functions (CDFs) of pitted and non-pitted fruit classes, it was possible to show a difference between these classes with fluorescence imaging. Results of pre-harvest imaging on apples to identify fruit with bitter pit potential at harvest showed that pitted fruit were correctly classified (75–100%). However, misclassification of non-pitted fruit (50% and less) with fluorescence imaging is still too high to be of any commercial.

Obenland & Neipp (2005) used Chl fluorescence analysis in green lemons *(Citrus union)* 30 minutes after immersion of the fruit into 55 C water for 5 minutes to determine if this methodology could be used to identify areas of hot water-induced rind injury before the appearance of visible symptoms. Fluorescence was variable in intensity over the surface of the rind with defined areas of enhanced fluorescence being present that corresponded in shape and location with visible injury that later developed during 24 hours of storage. The authors concluded that imaging of Chl fluorescence has potential as a means to identify areas of incipient rind injury in citrus to facilitate study of the causal mechanisms of postharvest rind disorders. On the other hand, previously Nedbal et al. (2000) demonstrated the potential for using rapid imaging of Chl fluorescence in post-harvest fruit to develop an automated device that can identify and remove poor quality fruit long before visible damage appears.

Meyerhoff & Pfündel (2008) used Chl fluorescence imaging to detect the presence of functioning PSII in fruits of strawberries. From obtained results authors concluded that it is unclear if photosynthesis in strawberry fruits is capable to support seed development.

Chl fluorescence imaging can be conveniently used to study the functioning of PSII in leaves and permits to detect the heterogeneity of photosynthesis which is particularly evident in stressed leaves. However, it has been reported as it can be conveniently used also for particular application such as the study of fruit quality in postharvest. For these reasons Chl fluorescence imaging represents an important and useful tool in ecophysiological and post harvest studies that permits to detect the effects of abiotic stress even at early stages and before the visual appearance of symptoms of damage.

# REFERENCES

1. M. Aldea, T. D. Frank, E. H. De Lucia, 2006 A method for quantitative analysis of spatially variable physiological processes across leaf surfaces. Photosynthesis Research, 90 161 172 , 0166-8595
2. J. M. Anderson, W. S. Chow, Y.I. Park, 1995 The grand design of photosynthesis: acclimation of the photosynthetic apparatus to environmental cues. Photosynthesis Research, 46 129 139 , 0166-8595
3. N. R. Baker, 2008 Chlorophyll Fluorescence: a probe of photosynthesis in vivo. Annual Review of Plant Biology, 59 89 113 , 1543-5008
4. N. R. Baker, T. M. East, S. P. Long, 1983 Chilling damage to photosynthesis in young Zea mays. II. Photochemial function of thylakoids in vivo. Journal of Experimental Botany, 34 189 197 , 0022-0957
5. N. R. Baker, E. Rosenqvist, 2004 Applications of chlorophyll fluorescence can improve crop production strategies: an examination of future possibilities. Journal of Experimental Botany, 55 1607 1621 , 0022-0957

6. S. Balachandran, C. B. Osmond, P. F. Daley, 1994 Diagnosis of the earliest strain-specific interactions between tobacco mosaic virus and chloroplasts of tobacco leaves in vivo by means of chlorophyll fluorescence imaging. Plant Physiology, 104 1059 1065 , 0032-0889
7. W. Bilger, O. Bjorkman, 1990 Role of the xanthophyll cycle in photoprotection elucidated by measurements of light-induced absorbance changes, fluorescence and photosynthesis in leaves of Hedera canariensis. Photosynthesis Research, 25 173 186 , 0166-8595
8. B. Biswal, U. C. Biswal, 1999 Photosynthesis under stress: stress signals and adaptive response of chloroplasts, In: Handbook of Plant and Crop Stress, Pessarakli, M. (ed), 315 336 , Marcel Dekker Inc., 0-82471-948-4 York
9. O. Björkman, B. Demmig, 1987 Photon yield of O2 evolution and chlorophyll fluorescence characteristics at 77K among vascular plants of diverse origins. Planta, 170 489 504 , 0032-0935
10. N. Boardman, 1977 Comparative photosynthesis of sun and shade plants. Annual Review of Plant Physiology, 28 355 377 , 0066-4294
11. J. M. Briantais, J. Dacosta, Y. Goulas, J. M. Ducruet, I. Moya, 1996 Heat stress induces in leaves an increase of the minimum level of chlorophyll fluorescence, F0- A time-resolved analysis. Photosynthesis Research, 48 189 196 , 0166-8595
12. E. Bro, S. Meyer, B. Genty, 1996 Heterogeneity of leaf CO2 assimilation during photosynthetic induction. Plant, Cell and Environment, 19 1349 1358 , 1365-3040
13. W. L. Butler, 1978 Energy distribution in the photochemical apparatus of photosynthesis. Annual Review of Plant Physiology, 29 345 378 , 0066-4294
14. A. Calatayud, E. Gorbe, D. Roca, P. F. Martinez, 2008 Effect of two nutrient solution temperatures on nitrate uptake, nitrate reductase activiy, NH4 + concentration and chlorophyll a fluorescence in rose plants. Environmental and Experimental Botany, 64 65 74 , 0098-8472
15. L. Chaerle, I. Leinonen, H. G. Jones, D. Van Der Straeten, 2006 Monitoring and screening plant population with combined

thermal and chlorophyll fluorescence imaging. Journal of Experimental Botany, 58 773 784 , 0022-0957

16. L. Chaerle, D. Van Der Straeten, 2000 Imaging techniques and the early detection of plant stress. Trends in Plant Science, 5 495 501 , 1360-1385
17. L. Chaerle, D. Van Der Straeten, 2001 Seeing is believing: imaging techniques to monitor plant health. Biochimica et Biophysica Acta, 1519 153 166 , 0304-4165
18. C. P. Chen, T. D. Frank, S. P. Long, 2009 Is a short, sharp shock equivalent to long-term punishment? Contrasting the spatial pattern of acute and chronic ozone damage to soybean leaves via chlorophyll fluorescence imaging. Plant Cell and Environment, 32 327 335 , 0140-7791
19. G. Cornic, C. Fresneau, 2002 Photosynthetic carbon reduction and oxidation cycles are the main electron sinks for photosystem II activity during a mild drought. Annals of Botany, 89 887 894 , 0305-7364
20. F. Dai, M. Zhou, X. , G. Zhang, P. , 2007 The change of chlorophyll fluorescence parameters in winter barley during recovery after freezing shock and as affected by cold acclimation and irradiance. Plant Physiology and Biochemistry, 45 915 921 , 0981-9428
21. J. R. De Ell, R. K. Prange, D. P. Murr, 1995 Chlorophyll fluorescence as a potential indicator of controlled-atmosphere disorders in Marshall Mcintosh apples. Hortscience, 30 1084 1085 0018-5345
22. J. R. De Ell, P. M. A. Toivonen, 2000 Chlorophyll fluorescence as a non-destructive indicator of broccoli quality during storage in modified atmosphere packaging. HortScience, 35 256 259 , 0018-5345
23. G. E. Edwards, N. R. Baker, 1993 Can assimilation in maize leaves be predicted accurately from chlorophyll fluorescence analysis? Photosynthesis Research, 37 89 102 0166-8595
24. B. Ehlert, D. K. Hincha, 2008 Chlorophyll fluorescence imaging accurately quantifies freezing damage and cold acclimation responses in Arabidopsis leaves. Plant Methods, 4, 12, 1746-4811 1746 4811

25. M. Fambrini, E. Degl'innocenti, G. Cionini, C. Pugliesi, L. Guidi, 2010 mesophyll cell defective1, a mutation that disrupts leaf mesophyll differentiation in sunflower. Photosynthetica, 48 135 142 , 0300-3604

26. Y.L. Feng, K.F. Cao, 2005 Photosynthesis and photoinhibition after night chilling in seedlings of two tropical tree species grown under three irradiances. Photosynthetica, 43 567 574 , 0300-3604

27. J. Flexas, M. Badger, W. S. Chow, H. Medrano, C. B. Osmond, 1999 Analysis of the relative increase in photosynthetic O2 uptake when photosynthesis in grapevine leaves is inhibited following low night temperatures and/or water stress. Plant Physiology, 121 675 684 , 0032-0889

28. J. Flexas, J. Bota, J. M. Escalona, B. Sampol, H. Medrano, 2002 Effects of drought on photosynthesis in grapevines under field conditions: an evaluation of stomatal and mesophyll limitations. Functional Plant Biology, 29 461 471 , 1445-4408

29. J. Flexas, J. Bota, J. Galmés, H. Medrano, M. Ribas-Carbó, 2006 Keeping a positive carbon balance under adverse conditions: responses of photosynthesis and respiration to water stress. Physiologia Plantarum, 127 343 352 , 0032-0889

30. J. Flexas, J. Bota, F. Loreto, G. Cornic, T. D. Sharkey, 2004 Diffusive and metabolic limitations to photosynthesis under drought and salinity in C3 plants. Plant Biology, 6 269 279 , 1435-8603

31. M. J. Fryer, J. R. Andrews, K. Oxborough, D. A. Blowers, N. R. Baker, 1998 Relationship between CO2 assimilation, photosynthetic electron transport and active O2 metabolism in leaves of maize in the field during periods of low temperature. Plant Physiology, 116 571 580 , 0032-0889

32. B. Genty, J. M. Briantais, N. R. Baker, 1989 The relationship between the quantum yield of photosynthetic electron transport and quenching of chlorophyll fluorescence. Biochimica et Biophysica Acta, 990 87 92 , 0304-4165

33. A. M. Gilmore, Govindjee, 1999 How higher plants respond to excess light: Energy dissipation in photosystem II, In: Concepts in Photobiology: Photosynthesis and Photomorphogenesis, Singhal G.S., Renger G., Irrgang K.D., Govindjee, Sopory S., 513 548 , Narosa Publishing House, New Delhi, 0-79235-519-9

34. G. Grassi, F. Magnani, 2005 Stomatal, mesophyll conductance and biochemical limitations to photosynthesis as affected by drought and leaf ontogeny in ash and oak trees. Plant, Cell and Environment, 28 834 849 , 0140-7791
35. L. Guidi, E. Degl'Innocenti, 2008 Ozone effects on high light-induced photoinhibition in Phaseolus vulgaris. Plant Science 174 590 596 , 0306-4484
36. L. Guidi, S. Mori, E. Degl'innocenti, S. Pecchia, 2007 Effects of ozone exposure or fungal pathogen on white lupin leaves as determined by imaging of chlorophyll a fluorescence. Plant Physiology and Biochemistry, 45 851 857 , 0981-9428
37. S. W. Hogewoning, J. Harbinson, 2007 Insights on the development, kinetics, and variation of photoinhibition using chlorophyll fluorescence imaging of a chilled, variegated leaf. Journal of Experimental Botany, 58 453 463 , 0022-0957
38. M. J. Hovenden, C. R. Warren, 1998 Photochemistry, energy dissipation and cold-hardening in Eucalyptus nitens and E. pauciflora. Australian Journal of Plant Physiology, 25 581 589 , 0310-7841
39. J. Kangasjarvi, P. Jaspers, H. Kollist, 2005 Signalling and cell death in ozone-exposed plants. KangasjarviJ.JaspersP.KollistH. (2005). Signalling and cell death in ozone-exposed plants. Plant and Cell Environmnet, , ISSN 0140-7791 and Cell Environmnet, 28 1021 1036 , 0140-7791
40. H. Kaustky, W. Appel, H. Amann, 1960 Chlorophyllfluorescenz und kholensaureassimilation. Biochemische Zeitschrift, 322 277 292 , 0366-0753
41. G. H. Krause, 1988 Photoinhibition of photosynthesis. An evaluation of damaging and protective mechanisms. Physiologia Plantarum, 74 566 574 , 0031-9317
42. J. Leipner, K. Oxborough, N. R. Baker, 2001 Primary sites of ozone-induced perturbations of photosynthesis in leaves: identification and characterization in Phaseolus vulgaris using high resolution chlorophyll fluorescence imaging. Journal of Experimental Botany, 52 1689 1696 , 0022-0957
43. H. K. Lichtenthaler, 1981 Adaptation of leaves and chloroplasts to high quanta fluence rate, In: Photosynthesis, Akoyunoglou G., 273 285 , Balaban International Science Service, 0-86689-012-2

44. H. K. Lichtenthaler, 1984 Differences in morphology and chemical composition of leaves grown at different light intensities and qualities, In: Control of leaf growth, S.E.B. Seminaries, Baker N.R., Davies W.J., Ong C.K., 201 221 , Cambridge University Press, 978-0-52110-362-6 Cambridge

45. H. K. Lichtenthaler, A. Ac, M. V. Marek, J. Kalina, O. Urban, 2007 Differences in pigment composition, photosynthetic rates and chlorophyll fluorescence images of sun and shade leaves of four tree species. Plant Physiology and Biochemistry, 45 577 588 , 0000-0981- 9428

46. S. P. Long, C. J. Bernacchi, 2003 Gas exchange measurements, what can they tell us about the underlying limitations to photosynthesis? Procedures and sources of error. Journal of Experimental Botany, 54 2393 2401 , 0022-0957

47. E. Lötze, C. Huybrechts, A. Sadie, K. I. Theron, R. M. Valcke, 2006 Fluorescence imaging as a non-destructive method for pre-harvest detection of bitter pit in apple fruit (Malus domestica Borkh.). Postharvest Biology and Technology, 40 287 294 , 0925-5214

48. A. Massacci, S. M. Nabiev, L. Pietrosanti, S. K. Nematov, T. N. Chernikova, K. Thor, J. Leipner, 2008 Response of the photosynthetic apparatus of cotton (Gossypium hirsutum) to the onset of drought stress under field conditions studied by gas-exchange analysis and chlorophyll fluorescence imaging. Plant Physiology and Biochemistry, 46 189 195 , 0000-0981- 9428

49. K. Maxwell, G. N. Johnson, 2000 Chlorophyll fluorescence: a practical guide. Journal of Experimental Botany, 51 659 668 , 0022-0957

50. Q. Meng, K. Siebke, P. Lippert, B. Baur, U. Mukherjee, E. Weis, 2001 Sink-source transition in tobacco leaves visualized using chlorophyll fluorescence imaging. New Phytologist, 151 585 596 , 0002-8646X

51. S. Meyer, B. Genty, 1999 Heterogeneous inhibition of photosynthesis over the leaf surface of Rosa rubiginosa L. during water stress and ABA-treatment: diffusive induced metabolic component. Planta, 210 126 131 , 0032-0935

52. S. Meyer, K. Saccardy-Adji, F. Rizza, B. Genty, 2001 Inhibition of photosynthesis by Colletotrichum lindemuthianum in bean

leaves determined by chlorophyll fluorescence imaging. Plant, Cell and Environment, 24 947 956 , 1365-3040

53. O. Meyerhoff, E. Pfündel, 2008 Photosynthesis in ripe strawberries (Fragaria × ananassa) recording by a MAXI IMAGING-PAM. PAM Application Notes, 1, 19- 20, (http://www.walz.com/downloads/pan/PAN07004.pdf)

54. L. Nedbal, J. Soukupova, J. Withmarsh, M. Trtilek, 2000 Postharvest imaging of chlorophyll fluorescence from lemons can be used to predict fruit quality. Photosynthetica, 38 571 579 , 0300-3604

55. L. Nedbal, J. Whitmarsh, 2004 Chlorophyll fluorescence imaging of leaves and fruits, In: Chlorophyll a Fluorescence. A Signature of Photosynthesis, Papageorgiou G.C. & Govindjee (Eds), 389 408 ; Springer, 140203217 Dordrecht

56. D. Obenland, P. Neipp, 2005 Chlorophyll fluorescence imaging allows early detection and localization of lemon rind injury following hot water treatment. HortScience, 40 1821 1823 , 0018-5345

57. B. Osmond, O. Schwartz, B. Gunning, 1999 Photoinhibitory printing on leaves, visualized by chlorophyll fluorescence imaging and confocal microscopy, is due to diminished fluorescence from grana. Australian Journal of Plant Physiology, 26 717 724 ., ISSN, 0310-7841

58. K. Oxborough, N. R. Baker, 1997 An instrument capable of imaging chlorophyll a fluorescence from intact leaves at very low irradiance and at cellular and subcellular levels. Plant, Cell and Environment, 20 1473 1483 , 1365-3040

59. A. Pavlovic, L. Slovakova, C. Pandolci, S. Mancuso, 2011 On the mechanism underlying photosynthetic limitation upon trigger hair irritation in the carnivorous plant Venus flytrap (Dionaea muscipula Ellis). Journal of Experimental Botany, 62 1991 2000 , 0022-0957

60. M. L. Perez-Bueno, M. Ciscato, Ven. M. vande, I. Garcia-Luque, R. Valcke, M. Baron, 2006 Imaging viral infection studies on Nicotaian benthamiana plants infected with the pepper mild mottle tobamovirus. Photosynthesis Research, 90 111 123 , 0166-8595

61. L. Qing-Ming, B.B. Liu, Y. Wu, Z.R. Zou, 2008 Interactive effects of drought stresses and elevated CO2 concentration on

photochemistry efficiency of cucumber seedlings. Journal of Integrative Plant Biology, 50 1307 1317 , 1672-9072

62. R. S. Quilliam, P. J. Swarbrick, J. D. Scholes, S. A. Rolfe, 2006 Imaging photosynthesis in wounded leaves of Arabidopsis thaliana. Journal of Experimental Botany, 57 55 69 , 0022-0957

63. J. Scharte, H. Schon, E. Weis, 2005 Photosynthesis and carbohydrate metabolism in tobacco leaves during an incompatible interaction with Phytophthora nicotianae. Plan Cell and Environment, 28 1421 1435 , 1365-3040

64. J. D. Scholes, S. A. Rolfe, 1996 Photosynthesis an localized regions of oat leaves infected with crown rust (Puccinia coronata). Quantitative imaging of chlorophyll fluorescence. Planta, 199 573 582 , ISSN, 0032-0943

65. U. Schreiber, W. Bilger, 1993 Progress in chlorophyll-fluorescence research: major developments during the past years in retrospect. Progress in Botany, 54 151 173 , 0340-4773

66. P. J. Schwarbrick, P. Schulze-Lefert, J. D. Scholes, 2006 The metabolic consequence of susceptibility and the activation of race-specific or broad-spectrum resistance pathways in barley leaves challenged with the powder mildew fungus Blumeria graminis. Plan Cell and Environment, 29 1061 1076 , 1365-3040

67. K. E. Sommerville, T. E. Gimeno, M. C. Ball, 2010 Primary nerve (vein) density influences spatial heterogeneity of photosynthetic response to drought in two Acacia species. Functional Plant Biology, 37 840 848 , 1445-4408

68. M. Strand, G. Öquist, 1985 Inhibition of photosynthesis by freezing temperatures and high light levels in cold-acclimated seedlings of Scots pine (Pinus sylvestris).- II. Effects on chlorophyll fluorescence at room temperature and 77 K. Physiologia Plantarum, 65 117 123 , 0031-9317

69. S. Takahashi, M. R. Badger, 2010 Photoprotection in plants: a new light on photosystem II damage. Trends in Plant Science, 16 53 60 , 1360-1385

70. O. van Kooten, J. F. H. Snel, 1990 The use of chlorophyll fluorescence nomenclature in plant stress physiology. Photosynthesis Research, 25 147 150 , 0166-8595

71. J. D. West, D. Peak, J. Q. Peterson, K. A. Mott, 2005 Dynamics of stomatal patches for a single surface of Xanthium strumarium L. Leaves observed with fluorescence and thermal images. Plant Cell and Environment, 28 633 641 , 1365-3040

72. N. S. Woo, M. R. Badger, B. J. Pogson, 2008 A rapid, non-invasive procedure for quantitative assessment of drought survival using chlorophyll fluorescence. Plant Methods, 4, 27, 1746-4811 1746 4811

73. D. L. Zuluaga, S. Ponzali, E. Loreti, C. Pucciariello, E. Degl'Innocenti, L. Guidi, A. Alpi, P. Perata, 2008 Arabidopsis thaliana MYB75/PAP1 transcription factor induces anthocyanin production in transgenic tomato plants. Functional Plant Biology, 35 606 618 , 1445-4408

Chapter 3

# An Online Database for Plant Image Analysis Software Tools

Guillaume Lobet1*, Xavier Draye[2] and Claire Périlleux[1]

[1]Laboratory of Plant Physiology, PhytoSYSTEMS, University of Liège, Boulevard du Rectorat 27, 4000 Liège, Belgium

[2]Earth and Life Institute, Université catholique de Louvain, Croix du Sud 2 L7.05.11, 1348, Louvain-la-Neuve, Belgium

# ABSTRACT

## Background

Recent years have seen an increase in methods for plant phenotyping using image analyses. These methods require new software solutions for data extraction and treatment. These solutions are instrumental in supporting various research pipelines, ranging from the localisation of cellular compounds to the quantification of tree canopies. However, due to the variety of existing tools and the lack of central repository, it is challenging for researchers to identify the software that is best suited for their research.

## Results

We present an online, manually curated, database referencing more than 90 plant image analysis software solutions. The website, plant-image-analysis.org, presents each software in a uniform and concise manner enabling users to identify the available solutions for their experimental needs. The website also enables user feedback, evaluations and new software submissions.

## Conclusions

The plant-image-analysis.org database provides an overview of existing plant image analysis software. The aim of such a toolbox is to help users to find solutions, and to provide developers a way to exchange and communicate about their work.


# BACKGROUND

Many fields of plant sciences, ranging from physiological studies to breeding programs, rely on linking genotypes and phenotypes. Thanks to the increasing development of genotyping techniques, plant scientists have been generating an incredibly large amount of genetic data. However, to effectively use this genetic information, it must be explicitly linked, together with environmental characterisation, to the corresponding phenotypic responses. Unfortunately, the lack of appropriate phenotyping tools and methods often hinders these studies, making the phenotyping step the bottleneck in many research programs [1].

Current phenotyping pipelines often rely on imaging techniques [2, 3]. Indeed, images have several key features that make them valuable for plant research: they are multidimensional in nature, contain several layers of information (e.g. shape and colors), allow a temporal decoupling of sampling and analysis and are prone to standardized and fully automated treatment. In addition, with the widespread adoption of simple imaging tools [4,5], with the appearance of more complex technologies [6-8] and with the increasing availability of powerful hardware, a majority of plant scientists use one or several imaging techniques in their research.

Computer scientists and plant biologists have been developing software solutions to handle imaging data [9]. Most of the time, these developments were bound to specific applications, for example the analysis of shoot meristem functioning [10] or the 3D reconstruction of entire root systems [11]. Unfortunately, the ever increasing number of available tools and the diversity of communication means within the scientific community, make it difficult for the non-specialists to find the most appropriate solution for their analyses. We therefore created a new online database referencing available plant image analysis software and allowing new tools deposit. This paper describes the database which aims to bridge the gap between software developers and users and help scientists to find the tools they need.

# CONSTRUCTION AND CONTENT

## General Structure

The plant image analysis database was built as a web-based repository and is freely available at the following address: www.plant-image-analysis.org [12]. The web interface was designed to enable scientists to quickly identify the right tools for their research (Figure 1). Users can navigate through the complete list of software solutions, query the database with keywords or browse it with pre-defined criteria such as the type of organ to analyse, the measurements to make, the desired automation level, the operating system or the license type. Users can stay informed using either RSS fluxes or following the database Twitter account (@plant_image).

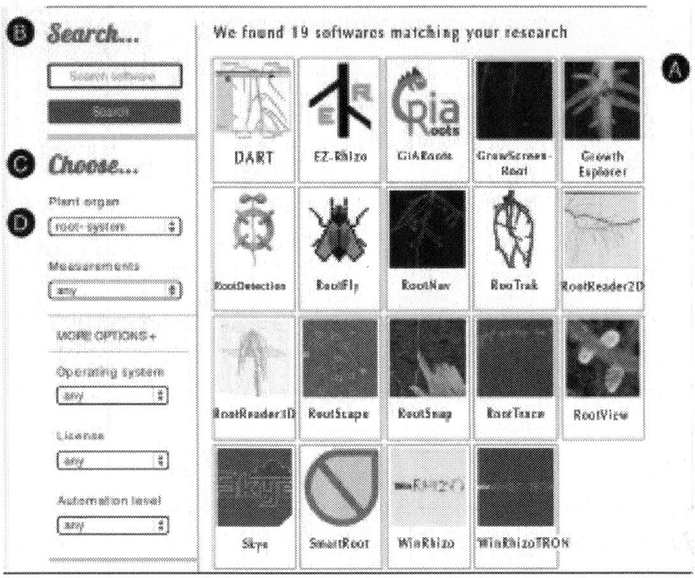


**Figure 1:** Main search page of the plant-image-analysis.org website.Users can browse through the software solutions (A), make a free search(B), or use pre-defined search criteria (C). Here, the list of software was restricted by the application of a filter on the organ type *root-system*(D).

## Data Sources and Management

Because the information about existing plant image analysis software is usually scattered among a variety of scientific publications (Figure 2) and websites, it is virtually impossible to automate data collection and database feeding. A manual curation of the database is therefore required to ensure that most of the available tools are presented and properly classified. The presence of broken links is periodically checked using Google Webmaster tools [13].

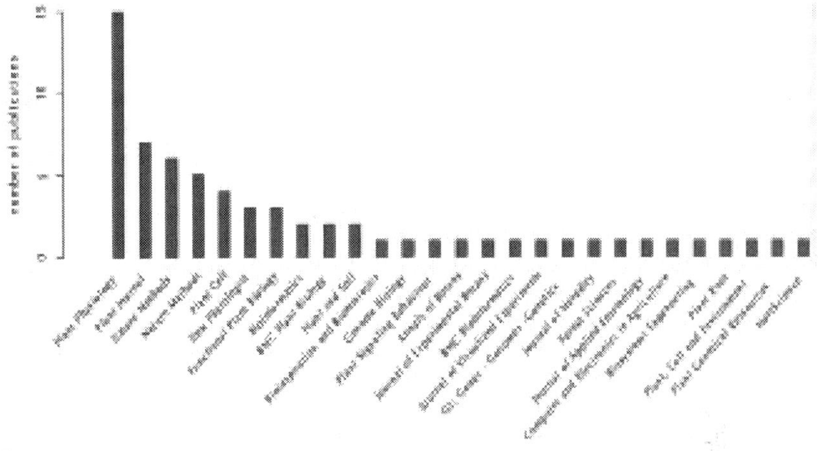

**Figure 2:** Number of publications presenting plant image analysis software per journal. Total number of journals = 27.

Both published and unpublished software were found through a thorough review of the literature and by using scientific reference databases (such as Scopus [14], ISI Web of Knowledge [15] or Google Scholar [16]) and regular search engines (such as Google). Although the list may not be exhaustive, we believe that most of the major plant image analysis software were found and incorporated. Up-dating the database is performed by automated web queries, literature screening (with a focus on journals where plant imaging tools were published, Figure 2) and developer contributions. Any software can be added on request by sending a predefined form to the database curators.

Regarding the database management, each software was assigned a set of keywords describing the target organ, measured parameters, automation level, license type and operating system. In order to avoid duplication and ambiguity, a limited number of keywords were selected. These keywords were chosen to describe the most basic features extracted from the images, excluding their combinations. As an example, for rosette analysis in Arabidopsis, "compactness" (defined as the ratio between rosette area and convex hull area, [17,18]) was not retained since it could be easily recalculated from its basic components. However, the number of keywords is not fixed and new features can be added if needed.

## Software Presentation and Description

The plant-image-analysis.org database is organized as a set of presentation sheets describing the different software in a concise and homogenized style (Figure 3). Firstly, a short description introduces each software, generally based on the information provided by its developer. Secondly, a formatted list of the main software features is reported including the plant part for which it was designed, the nature of the collected data, the level of automation, the operating system, the license type, the plant and image requirements and the export format. This set of information was chosen to meet the criteria usually used by researchers when searching for a software solution. References of related publications are given (if any), as well as the name of the developers.

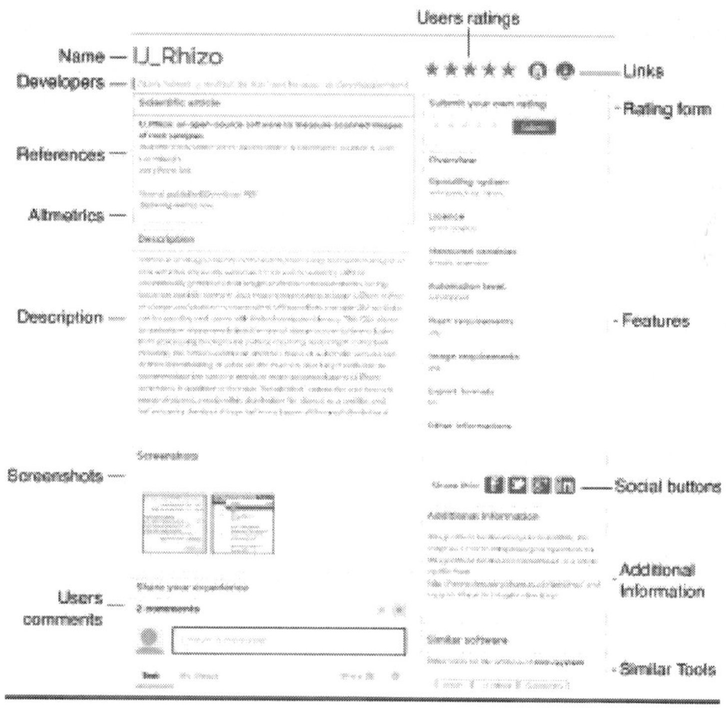


**Figure 3:** Software presentation page on the plant-image-analysis.org website. Software presentation page on the plant-image-analysis.org website (in this case, IJ-Rhizo's presentation page [19]).

In addition, taking advantage of recent web technologies, user feedback and software evaluations have been implemented in every presentation page. This layer of information, fed by the scientists themselves, aims at improving the tools and establishing interactions between software users and developers. Social media links (such as Twitter or Facebook) have been added to promote sharing and discussions [20]. Finally, altmetrics (alternative metrics [21]) have been added for every published article using a widget developed by ImpactStory [22]. These metrics convey information on the articles reception by the scientific community and provide the users with an indication of the relevancy of the paper in their field [23].

# UTILITY AND DISCUSSION

Currently, the database references more than 90 software, ranging from cell to whole canopy analyses. It provides a way for developers of plant image analysis tools to present their work, published or unpublished. It enables plant scientists to easily find and compare the different tools available for their research. Since its release in January 2013, the website received an average of 500 visitors per month (data retrieved from Google Analytics [24]), confirming the interest of the scientific community for such a repository.

## An Overview of the Available Plant Image Analysis Software

The plant-image-analysis.org database allows to draw an overview of the available plant image analysis tools (Figure 4).

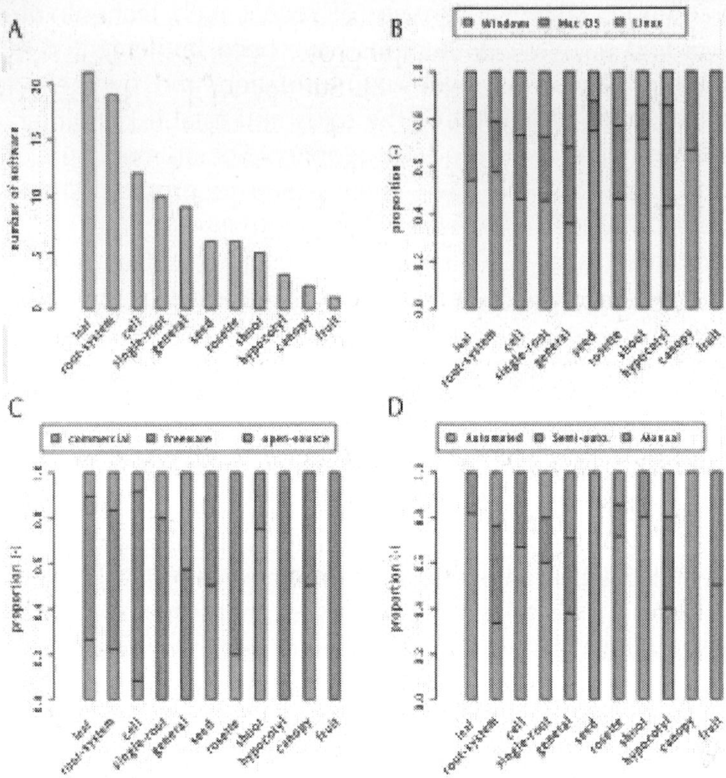


**Figure 4:** Distribution of the tools presented in the plant-image-analysis.org website. Distribution of the tools presented in the plant-image-analysis.org website, by plant organ type. A. Number of software by plant organ type. B. Proportion of operating systems by organ type. C. Proportion of license type by organ type. D. Proportion of automation levels by organ type. Total number of software = 91.

A first general observation is that the number of available systems varies very much with plant organs (Figure 4A). In particular, a large proportion of the tools are dedicated to individual leaves, then to the analysis of roots (either root systems or single roots) and cells. As stated earlier, the important number of similar, yet different, tools reflects the diversity of biological questions and hardware solutions. Software solutions are frequently developed in a specific context, making them unsuitable for other applications. Fortunately, developers have begun to address this issue as demonstrated by the recent publication of more flexible software solutions. As an example, for the analysis of rosette of Arabidopsis plants, Rosette Tracker provides a framework adaptable to multiple experimental designs [17]. Still for the analysis of rosette (or individual leaves), PhenoPhyte is accessible through a web-interface, making it a easy cross-platform solution that does not require any installation on a local machine [25]. For the analysis of root system, both SmartRoot [26] and RootNav [27] rely on semi-automated root tracing procedures that make them suitable for a large range of image types and qualities thanks to their semi-automated root tracing process.

While many plant image analysis applications were originally developed for the Windows operating system (Figure 4B), it should be noted that many developers have made efforts to offer cross-platform solutions. Moreover, a large majority of the referenced software are available for free for the scientific community (and even open-source, Figure 4C). Both the inter-operability between operating systems and the free access highlight positive dynamics in the on-going development of new plant image analysis software.

Finally, in line with the growing development of phenotyping platforms and pipelines, a clear tendency towards a full automation of the image analyses process can be observed (Figure 4D). Interestingly, for few organ types such as the root systems, a fair proportion of manual or semi-automated tools are available. In the case of the analysis of root systems, this distribution can be explained, at least partially, by the intrinsic nature of the objects to analyse. Root systems are indeed highly branched and complex structures, which makes automated analysis challenging and prone to cumulative errors [27]. In such scenario, semi-automated and manual procedures are still needed for the acquisition of quality data.

Our analysis shows that, in many cases, new users will have the choice between different software solutions (Figure 5A) that were developed independently for specific purposes and with given constrains. Although these tools present some redundancy (Figure 5B), they also contain complementarity approaches that could be combined for further improvement. By providing a classified and homogenised presentation of the available image analysis software solutions, the plant-image-analysis.org database aims at increasing the communication between developers, so participating to the concerted development of future tools.

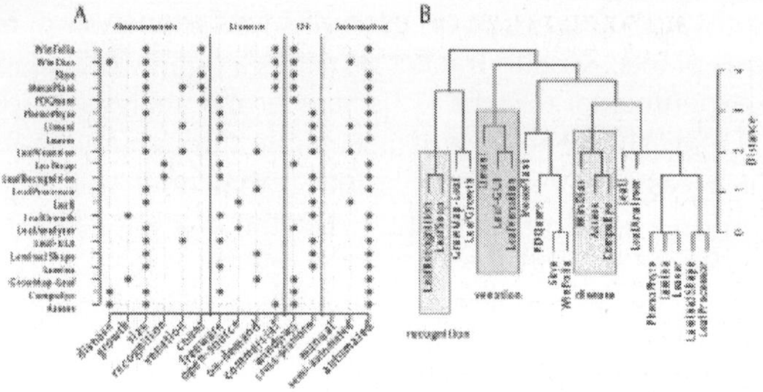

**Figure 5:** Illustration of the redundancy between the existing plant image analysis software solutions. Illustration of the redundancy between the existing plant image analysis software solutions, here for the analysis of single leaves. A. Properties of each software (measurements, license, operating system and automation level). B. Clustering of the different tools, based on their properties. A higher weight was given to the measurements compared to the other properties for the clustering (2:1). Tools designed for specific measurements (disease, venation analysis or species recognition) are highlighted in colors. For both figures, and for the sake of clarity, measurements were pooled by classes (e.g. *size* encompasses surface, length, width and perimeter measurements).

# CONCLUSION

The variety of biological questions, hardware solutions and technical approaches in plant image analysis have led to the development of a wide variety of tools and software. The diversity of hosting solution (from personal webpages to centralized repositories) and publication type (from none to biological to computational journals) has led to the dispersion of these tools across the web, making it difficult for a researcher to find the right tool for his research.

Here we presented a new online, manually curated, database that references and presents more than 90 plant image analysis tools. This database enables developers to present their tools (both published or unpublished) and users to easily navigate through the space of possible software solutions to find the most appropriate solution for their research.

# ACKNOWLEDGEMENTS

This research was supported by the Interuniversity Attraction Poles Program P7/29 (MARS project) initiated by the Belgian Science Policy Office, a grant from the Fonds pour la formation à la Recherche dans l'Industrie et dans l'Agriculture and the European Community's Seventh Framework Programme under the grant agreement FP7-244374 (DROPS project).

The authors would like to thanks Dr. Pierre Tocquin (ULg) for his critical reading of the manuscript and insightful comments regarding the database and Dr. Michael Pound (CPIB) for improving the English of the manuscript.


# REFERENCES


1. Cobb JN, DeClerck G, Greenberg A, Clark R, McCouch S: Next-generation phenotyping: requirements and strategies for enhancing our understanding of genotype-phenotype relationships and its relevance to crop improvement. *TAG Theor Appl Genet* 2013, 126:867-887.
2. Dhondt S, Wuyts N, Inzé D: Cell to whole-plant phenotyping: the best is yet to come. *TRENDS in Plant Sci* 2013, 18(8):428-439.
3. Spalding EP, Miller ND: Image analysis is driving a renaissance in growth measurement. *Curr Opin Plant Biol* 2013, 16:100-104.
4. Lobet G, Draye X: Novel scanning procedure enabling the vectorization of entire rhizotron-grown root systems. *Plant Methods* 2013, 9(1):2-10.
5. Neumann G, George TS, Plassard C: Strategies and methods for studying the rhizosphere - the plant science toolbox. *Plant Soil* 2009, 321:431-456.
6. Esser HG, Carminati A, Vontobel P, Lehmann EH, Oswald SE: Neutron radiography and tomography of water distribution in the root zone. *J Plant Nutrition Soil Sci* 2010, 173(5):757-764.
7. Mooney SJ, Pridmore TP, Helliwell J, Bennett MJ: Developing X-ray Computed Tomography to non-invasively image 3-D root systems architecture in soil. *Plant Soil* 2012, 352(1):1-22.

8. Downie H, Holden N, Otten W, Spiers AJ, Valentine TA, Dupuy L: Transparent soil for imaging the rhizosphere. *PLoS ONE* 2012, 7(9):1-6.
9. Pridmore TP, French AP, Pound MP: What lies beneath: underlying assumptions in bioimage analysis. *TRENDS Plant Sci* 2012, 17(12):688-692.
10. Fernandez R, Das P, Mirabet V, Moscardi E, Traas J, Verdeil JL, Malandain G, Godin C:Imaging plant growth in 4d: robust tissue reconstruction and lineaging at cell resolution. *Nat Methods* 2010, 7(7):547-553.
11. Mairhofer S, Zappala S, Tracy S, Sturrock C, Bennett MJ, Mooney SJ, Pridmore TP:Recovering complete plant root system architectures from soil via X-ray mu-Computed Tomography. *Plant Methods* 2013, 9(8):1-7.
12. Lobet G: Plant Image Analysis. [http://www.plant-image-analysis.org Google Webmaster Tools [https://support.google.com/webmasters Scopus [http://www.scopus.com
13. ISI Web of Knowledge [http://www.isiknowledge.com
14. Google Scholar [http://scholar.google.be/
15. De Vylder J, Vandenbussche F, Hu Y, Philips W, Van Der Straeten D: Rosette Tracker: an open source image analysis tool for automatic quantification of genotype effects. *Plant Physiol* 2012, 160(3):1149-1159.
16. Arvidsson S, Pérez-Rodríguez P, Mueller-Roeber B: A growth phenotyping pipeline for Arabidopsis thaliana integrating image analysis and rosette area modeling for robust quantification of genotype effects. *New Phytologist* 2011, 191(3):895-907.
17. Pierret A, Gonkhamdee S, Jourdan C, Maeght JL: IJ-Rhizo: an open-source software to measure scanned images of root samples. *Plant Soil* 2013, 1-9. in press
18. Osterrieder A: The value and use of social media as communication tool in the plant sciences. *Plant Methods* 2013, 9(1):26.
19. Priem J, Taraborelli D, Groth P, Neylon C: Altmetrics: A Manifesto, (v.1.0).[http://www.altmetrics.org
20. Priem J, Piwowar H: Impact Story. [http://www.impactstory.org
21. Priem J: Beyond the paper. *Nature* 2013, 495:437-440.

22. Google Analytics [http://www.google.com/analytics/
23. Green JM, Appel H, MacNeal Rehrig E, Harnsomburana J, Chang JF, Balint-Kurti P, Shyu CR:PhenoPhyte: a flexible affordable method to quantify 2D phenotypes from imagery. *Plant Methods* 2012, 8(45):1-12.
24. Lobet G, Pagès L, Draye X: A novel image analysis toolbox enabling quantitative analysis of root system architecture. *Plant Physiol* 2011, 157(1):29-39.
25. Pound MP, French AP, Atkinson J, Wells DM, Bennet JM, Pridmore T: RootNav: Navigating images of complex root architectures. *Plant Physiol* 2013, 162(4):1802-1814.

Chapter 4

# Estimating Rice Chlorophyll Content and Leaf Nitrogen Concentration with a Digital Still Color Camera under Natural Light

Yuan Wang[1], Dejian Wang[1*], Peihua Shi[2] and Kenji Omasa[3]

[1]Institute of Soil Science, Chinese Academy of Sciences, 71 East Beijing Road, Nanjing 210008, PR China

[2]National Engineering and Technology Center for Information Agriculture, Nanjing Agricultural University, 1 Weigang Road, Nanjing 210095, PR China

[3]Graduate School of Agricultural and Life Sciences, The University of Tokyo, Yayoi 1-1-1, Bunkyo-ku, Tokyo 113-8657, Japan

# ABSTRACT

## Background

The color of crop leaves is closely correlated with nitrogen (N) status and can be quantified easily with a digital still color camera and image processing software. The establishment of the relationship between image color indices and N status under natural light is important for crop monitoring and N diagnosis in the field. In our study, a digital still color camera was used to take pictures of the canopies of 6 rice (*Oryza sativa* L.) cultivars with N treatments ranging from 0 to 315 kg N $ha^{-1}$ in the field under sunny and overcast conditions in 2010 and 2011, respectively.

## Results

Significant correlations were observed between SPAD readings, leaf N concentration (LNC) and 13 image color indices calculated from digital camera images using three color models: RGB, widely used additive color model; HSV, a cylindrical-coordinate similar to the human perception of colors; and the$L^*a^*b^*$ system of the International Commission on Illumination. Among these color indices, the index$b^*$, which represents the visual perception of yellow-blue chroma, has the closest linear relationship with SPAD reading and LNC. However, the relationships between LNC and color indices were affected by the developmental phase. Linear regression models were used to predict LNC and SPAD from color indices and phasic development. After that, the models were validated with independent data. Generally, acceptable performance and prediction were found between the color index $b^*$, SPAD reading and LNC with different cultivars and sampling dates under different natural light conditions.

## Conclusions

Our study showed that digital color image analysis could be a simple method of assessing rice N status under natural light conditions for different cultivars and different developmental stages.

# BACKGROUND

Nitrogen (N) is one of the most important nutrients essential for the growth of crops, and is a major component of chlorophyll and protein which are closely associated with leaf color, crop growth status and yield [1]. Insufficient N supply leads to smaller leaves, lower chlorophyll content and less biomass production, and consequently, reduced grain yield and quality [2,3]. Conversely, excessive N application can lead to the environmental problems of water and atmospheric pollution [4]. Hence, measuring crop N status timely is critical for increasing N use efficiency and environmental quality [5,6].

The spectral reflectance of crop leaf or canopy is known to be correlated with N status [7,8]. The instruments for measuring spectral reflectance are the chlorophyll meter [9-11], multi-spectral sensor [12], hyper-spectral sensor [7,8] and commercial digital camera [13-15], are used in precision agriculture for growth monitoring, nitrogen diagnosis and site-specific crop management. The SPAD meter (SPAD-502, Minolta Camera Co., Osaka, Japan), measures leaf chlorophyll content nondestructively, has potential for improving N use efficiency without affecting grain yield in real-time nitrogen management (RTNM) experiments [9,11]. A disadvantage of the SPAD-502 for assessing crop N status is its small sampling area (6 $mm^2$). In addition, the measurements are subject to operator bias so a large number of repetitions are needed to obtain reliable results [16,17], and SPAD meter experiences difficulties in distinguishing chlorophyll levels when crops are near or above the optimum N supply [10,18]. In contrast, satellite or airborne-mounted hyperspectral sensors can obtain spectral information in a larger sampling area and record more spectral bands [7,8,12]. The high cost of images, infrequent satellite overpasses and risk of images being obscured by clouds limit the application of these platforms for commercial use [19]. The proximal sensors GreenSeeker® (NTech Industries Inc., USA) and Yara N-sensor (Yara International ASA, Germany) which measure red and near infra-red (NIR) reflectance, overcome some of the limitations of satellite or airborne sensors [20], but their accuracy is influenced by background soil interference [8].

Alternatively, images from digital still color cameras, which record spectral information of visible bands, have a low cost but very high image resolution (consumer cameras in 2014 record up to 40 megapixels per image). At a sensor height of about 1 meter above the canopy, high-spatial-resolution images separate crops from background soil or other interferences, which is important for accurate diagnosis of N status when the vegetation fraction was low [13,21]. Moreover, images from digital still cameras contain a large amount of information about the crop structure and leaf color, such as leaf orientation, plant height, biomass accumulation and leaf senescence [22-24], and these parameters are easy to obtain [13,25] with existing software, such as MatLab® (MathWorks Inc.), the free-ware package ImageJ [26]. Previous studies showed that canopy cover estimated from the images was not only highly correlated with leaf area index (LAI), aboveground biomass and N accumulation [13,27], but was also stable in varying environmental conditions [28,29]. In addition to canopy cover, color digital images provide spectral information in the visible bands which are closely related to the leaf N concentration (LNC) and SPAD readings [17]. Hunt *et al.*[30] found that the triangular greenness index (TGI), which was derived from red, green and blue bands of a digital still color camera, was sensitive to leaf chlorophyll content of a whole canopy.

The color-related indices from digital still cameras can diagnose crop N status [17,30,31]. Previous studies on the analysis of leaf color were mostly undertaken in controlled light conditions [17,32-34]. This approach could reduce the impact of light on the image color, and easily obtain a reliable relationship between N status and leaf color indices. However, the results derived from the controlled light cannot be completely applied to natural light because of the variable light conditions [15,35,36]. Besides, there are still many uncertainties in the use of digital still cameras for N diagnosis under natural light conditions, and further validation is necessary to ensure the application in the field.

In this study, experiments with different N application rates were carried out in the field under sunny and overcast conditions (1) to analyze the relationship between chlorophyll content, LNC and canopy color related indices in different cultivars and stage of phasic development, (2) to establish the possible models for the diagnosis of crop N status using image color indices, and (3) to validate the applicability of the models under different natural light conditions.

# RESULTS AND DISCUSSION

## Correlation of Color Related Indices and Crop Nitrogen Status

Correlations between two rice N parameters (leaf N concentration and SPAD reading) and 13 image-color related indices (Eq. 1–10) were calculated with individual and pooled cultivars in 2011. Similar results were obtained from the three rice cultivars, i.e., Liangyoupeijiu, Nanjing45 and Nanjing46. Therefore, only correlation coefficients from the Liangyoupeijiu dataset (n = 72) and the pooled dataset (n = 240) were displayed in Table 1. SPAD readings showed significant correlations with each color index except g in both datasets. Among these color indices, *L**, *b**, R, G, B, r and INT were negatively correlated with SPAD readings, while the other indices were positively correlated with SPAD readings. Most of the color indices were closely correlated with LNC, however, the magnitude and direction of the correlation coefficients were not consistent with those between color indices and SPAD readings. Overall, the indices derived from the CIE *L*a*b** color model had relatively higher correlation coefficients with SPAD readings and LNC. In this color model, the index *b** represents the visual perception of yellow-blue chroma which is similar with the leaf color variation, and it has been used in many other studies for image color analysis [37,38]. Therefore, we select index *b** as a representative for further analysis.

**Table 1:** Correlation coefficients between SPAD readings, leaf nitrogen concentration (LNC, g kg-1) and image color related indices (digital number from three color models: RGB, HSV and CIE*L*a*b**)

| | SPAD | LNC | L* | a* | b* | b*/a* | H | R | G | B | r | g | b | INT | |
|---|---|---|---|---|---|---|---|---|---|---|---|---|---|---|---|
| SPAD | 1.00 | 0.21 | -0.67** | 0.78** | -0.90** | 0.76** | 0.83** | -0.68** | -0.67** | -0.48** | -0.76** | 0.01 | 0.41** | -0.63** | 0.66** |
| LNC | 0.32** | 1.00 | 0.48** | -0.26* | -0.06 | 0.35** | 0.13 | 0.47** | 0.48** | 0.63** | -0.37** | -0.68** | 0.65** | 0.52** | -0.28* |
| L* | -0.62** | 0.35** | 1.00 | -0.83** | 0.73** | -0.43** | -0.63** | 1.00** | 1.00** | 0.96** | 0.38** | -0.64** | 0.21 | 1.00** | -0.90** |
| a* | 0.61** | -0.04 | -0.70** | 1.00 | -0.86** | 0.50** | 0.69** | -0.82** | -0.84** | -0.69** | -0.52** | 0.18 | 0.17 | -0.80** | 0.62** |
| b* | -0.73** | -0.35** | 0.54** | -0.72** | 1.00 | -0.86** | -0.95** | 0.74** | 0.73** | 0.51** | 0.87** | 0.04 | -0.51** | 0.68** | -0.73** |
| b*/a* | 0.60** | 0.48** | -0.29** | 0.36** | -0.91** | 1.00 | 0.96** | -0.45** | -0.43** | -0.18 | -0.99** | -0.26* | 0.72** | -0.37** | 0.63** |
| H | 0.61** | 0.40** | -0.39** | 0.42** | -0.91** | 0.97** | 1.00 | -0.64** | -0.63** | -0.40** | -0.95** | -0.09 | 0.58** | -0.58** | 0.75** |
| R | -0.63** | 0.33** | 1.00** | -0.68** | 0.56** | -0.34** | -0.44** | 1.00 | 1.00** | 0.96** | 0.39** | -0.64** | 0.21 | 1.00** | -0.92** |
| G | -0.62** | 0.35** | 1.00** | -0.71** | 0.54** | -0.29** | -0.39** | 1.00** | 1.00 | 0.96** | 0.38** | -0.64** | 0.21 | 1.00** | -0.90** |
| B | -0.37** | 0.58** | 0.91** | -0.47** | 0.15* | 0.09 | -0.03 | 0.90** | 0.91** | 1.00 | 0.12 | -0.82** | 0.48** | 0.98** | -0.83** |
| r | -0.56** | -0.54** | 0.19** | -0.35** | 0.89** | -0.99** | -0.95** | 0.24** | 0.19** | -0.20** | 1.00 | 0.36** | -0.79** | 0.31** | -0.56** |
| g | 0.05 | -0.66** | -0.62** | 0.03 | 0.31** | -0.44** | -0.34** | -0.60** | -0.61** | -0.87** | 0.56** | 1.00 | -0.86** | -0.69** | 0.58** |
| b | 0.27** | 0.69** | 0.27** | 0.17** | -0.66** | 0.79** | 0.71** | 0.23** | 0.26** | 0.63** | -0.87** | -0.90** | 1.00 | 0.29* | -0.08 |
| INT | -0.56** | 0.42** | 0.99** | -0.64** | 0.45** | -0.20** | -0.31** | 0.99** | 0.99** | 0.95** | 0.10 | -0.70** | 0.37** | 1.00 | -0.89** |
| $VI_{Green}$ | 0.67** | -0.10 | -0.85** | 0.42** | -0.65** | 0.62** | 0.68** | -0.88** | -0.85** | -0.68** | -0.51** | 0.43** | 0.02 | -0.83** | 1.00 |

Numbers in the upper diagonal were calculated with the Liangyoupeijiu (n = 72) dataset and numbers in the lower diagonal were calculated with the combined data of Liangyoupeijiu, Nanjing45 and Nanjing46 in 2011 (n = 240).

**indicate the significance at $P < 0.01$, *indicate the significance at $P < 0.05$.

## Relationships between SPAD, LNC and the Color Index *b**

Regression analyses were performed between SPAD readings, LNC and the color index *b** using the 2011 dataset. Positive linear relationships were observed between LNC and SPAD readings, with the same trends in different sampling dates and cultivars (Figure 1). The determination coefficient ($R^2$) in different sampling dates varied from 0.61 to 0.88 along with the root mean square error (RMSE) from 1.81 to 2.64 g $kg^{-1}$. The LNC decreased with the rice development, while the maximum SPAD values increased with rice growth. Smaller RMSE was obtained in jointing and booting stages than in vegetative and tillering stages. Similarly, Xue et al. [39] reported that the ratio index of NIR/green ($R_{810}/R_{560}$) reached the best accuracy with LNC at jointing stage. When data were pooled across the sampling dates, there was no significant trend observed between SPAD readings and LNC (Figure 1e).

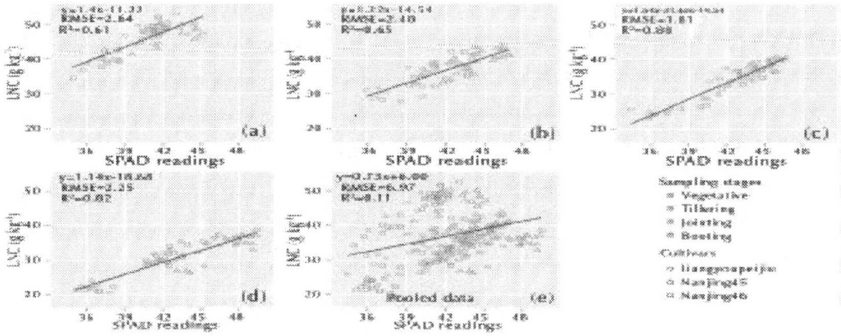


**Figure 1:** Relationships between leaf nitrogen concentration (LNC, g $kg^{-1}$) and SPAD readings in vegetative (a), tillering (b), jointing (c), booting (d) stages, and the pooled data of the four stages (e) in 2011.

Significant negative linear relationships were seen between LNC and color index *b**, with $R^2$ ranging from 0.58 to 0.86 in the four developmental stages (Figure 2). The $R^2$ between *b** and LNC at booting stage (Figure 2d) was less than that between SPAD readings and LNC. Same as the relationship between SPAD readings and LNC, no significant trend was observed between *b** and LNC when data pooled across the sampling dates (Figure 2e). In addition, there were negative linear relationships between *b** and SPAD readings (Figure 3). The $R^2$ in the vegetative stage was lower than that in the other stages. There were obvious differences among cultivars for the relationship between *b** and SPAD readings, especially the Liangyoupeijiu in jointing and booting stages (Figure 3c and d) and the Nanjing45 in vegetative stage (Figure 3a). In this case, the regression analysis was carried out with individual cultivars (Table 2). Overall, higher $R^2$ were observed from individual cultivars and sampling dates than that from the pooled dataset (Table 2). However, the intercepts and slopes of the linear relationship varied with rice cultivars, these differences might be partly caused by the different plant type among cultivars, with a loose shape and large mean leaf angle in hybrid indica rice (Liangyoupeijiu) [40,41] while tight shape and small mean leaf angle in japonica rice (Nanjing45, Nanjing46). The different plant type lead to different distribution of reflectance [42, 43] and finally caused different image-derived indices.

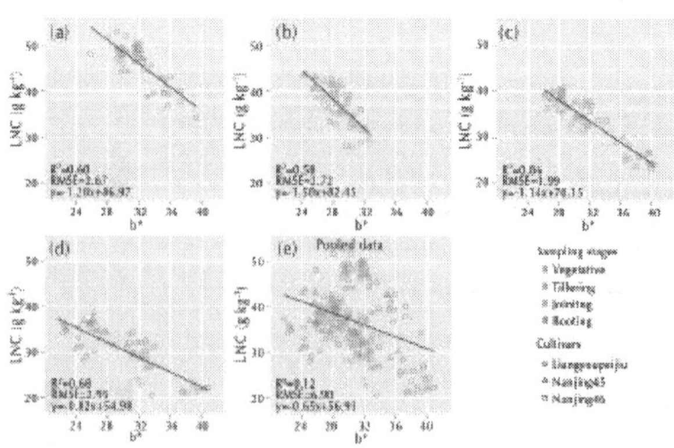


**Figure 2:** Relationships between leaf nitrogen concentration (LNC, g kg$^{-1}$) and image color index*b**in vegetative (a), tillering (b), jointing (c), booting (d) stages, and the pooled data of the four stages (e) in 2011.

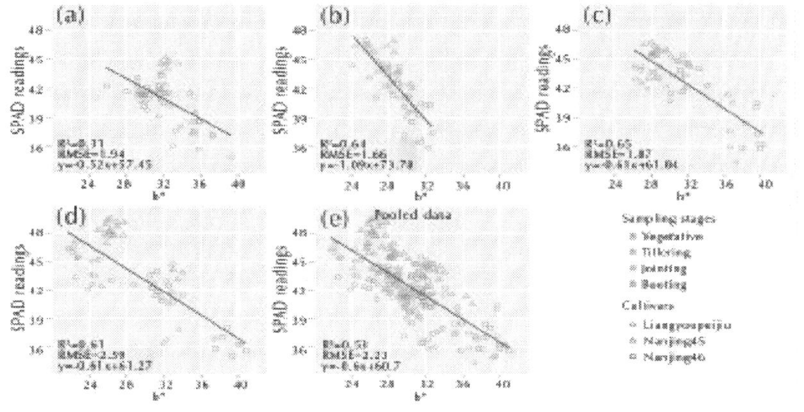


**Figure 3**: Relationships between SPAD readings and image color index*b**in vegetative (a), tillering (b), jointing (c), booting (d) stages, and the pooled data of the four stages (e) in 2011.

**Table 2:** Statistics of the linear regression analysis between color index*b**and SPAD readings in different development stages and cultivars in 2011

| | | Vegetative | Tillering | Jointing | | |
|---|---|---|---|---|---|---|
| Liangyoupeijiu | $R^2$ | 0.74** | 0.62** | 0.92** | 0.86** | 0.80** |
| | RMSE[a] | 0.87 | 1.18 | 0.86 | 1.23 | 1.24 |
| Nanjing45 | $R^2$ | 0.42** | 0.28** | 0.90** | 0.79** | 0.60** |
| | RMSE | 1.61 | 1.41 | 0.64 | 0.91 | 1.60 |
| Nanjing46 | $R^2$ | 0.83** | 0.52** | 0.82** | 0.83** | 0.59** |
| | RMSE | 0.94 | 1.62 | 1.47 | 1.26 | 1.85 |
| All cultivars | $R^2$ | 0.31** | 0.64** | 0.65** | 0.61** | 0.53** |
| | RMSE | 1.94 | 1.66 | 1.87 | 2.59 | 2.23 |

Dependent variable: SPAD.

**indicate the significance at $P < 0.01$.

[a]Root mean square error.

In our experiments, image acquisition was carried out in the field under natural light near solar noon which was the period with the most stable illumination at the top of the atmosphere. This makes sure that the light intensity is not changing too much during image acquisition in a single day. The results from Table 2indicated that reliable estimates of N status could be obtained from images taken under natural light. Considering the stability of N diagnosis in different locations and sampling dates, individual sampling dates could not meet the needs of crop monitoring and N diagnosis in various environmental conditions. Therefore, regression analysis was carried out with the pooled data of sampling dates. There were large differences of light intensity among different sampling dates even under overcast days (Table 3, PAR ranging from 145 to 692 $\mu$mol m$^{-2}$ s$^{-1}$ in 2011).

**Table 3:** Rice cultivars, sampling dates (indicated as days after transplanting, DAT), photosynthetic active radiation (PAR,$\mu$mol m$^{-2}$ s$^{-1}$) and the number of samples in the two experiments

| Experi-ment | Year | Cultivar | Vegetative | Tillering | Jointing | Booting | Number of samples |
|---|---|---|---|---|---|---|---|
| I | 2010 | Wuyunjing24 | 25[a] (937)[b] | 36 (1536) | 50 (1369) | 64 (1532) | 18 |
| | 2011 | Nanjing46 | 18 (145) | 35 (692) | 55 (203) | 75 (296) | 24 |
| II | 2010 | Nanjing44 | 22 (1215) | 35 (1058) | 49 (1759) | 63 (1477) | 18 |
| | | Yangjing48 | 22 (1449) | 35 (1058) | 49 (1759) | 63 (1477) | 18 |
| | 2011 | Nanjing45 | 16 (427) | 29 (654) | 51[c] (589) | 57 (289) | 18 |
| | | Liangyoupeijiu | 16 (427) | 29 (654) | 51 (589) | 65 (621) | 18 |

[a]represents sampling dates (the days after transplanting, DAT).

[b]represents the average PAR during the period of image acquisition.

[c]Because of the continuous sunny days in the jointing stage for Nanjing45, sampling dates was delayed about a week.

Interestingly, regression analysis did not show any evidence that the relationship between SPAD and color index $b^*$ was affected by the varying light intensity (Figure 3e). This might be attributed to the auto exposure controlled by the digital camera which adjusted the exposure time to make compensation for the amount of light reaching the image sensor. However, there were no significant trends between LNC and SPAD readings, or between LNC and color index $b^*$using data pooled across different sampling dates (Figures 1e and 2e). Previous studies revealed that, for rice and corn, the relationship between LNC and

SPAD readings could be improved simply by dividing the readings with specific leaf weight (SLW = dry leaf weight/leaf area) of the sampled leaves or introducing SLW as a second independent variable in the multiple regression [44-46]. The reason is that SPAD readings vary with leaf thickness which can be different in cultivars, developmental stages and environmental conditions [46,47], while the LNC has a relatively consistent value. The uncertain relationship between LNC and color index $b^*$ (Figure 2e) may also be caused by the difference of leaf thickness, because the color index $b^*$ and SPAD readings both reveal the spectral information of leaves, and their relationship keeps consistent with the pooled data of different sampling dates (Table 2 and Figure 3e). Nowadays, destructive sampling or hyperspectral-reflectance [48] is required for the measurement of SLW, however, including this defeats the purpose of using a cheap and simple digital camera.

## Model Calibration and Validation

Since developmental stages in rice affected the response of N parameters to color indices, multiple linear regression analysis was carried out with the consideration of days after transplanting (DAT) to estimate LNC and SPAD (Table 4). The multiple linear models highly improved the $R^2$ of SPAD readings and $b^*$with LNC (Table 4, Figure 4a and b). The relationships between SPAD readings and $b^*$ were less affected by the process of development for all the cultivars (Table 2 and Figure 3e). Therefore, the consideration of DAT in the multiple linear regression did not improve the $R^2$ effectively (Figure 4c). The slope of the regression lines in Figure 4 was all less than 1, which indicated that predicted LNC or SPAD were generally smaller at the high value area and bigger at the low value area, than the observed ones. It was noteworthy that most of the predicted LNC in the vegetative stage (red symbols) were underestimated in Figure 4a and b. This is mainly because that the plants at this stage has higher concentration of N but lower concentration of chlorophyll. In this case, the lower concentration of chlorophyll would cause an underestimation of the LNC in the vegetative stage.

**Table 4:** Statistics of the calibration and validation results for estimating leaf nitrogen concentration (LNC, g kg-1) and SPAD with color index*b**, SPAD and days after transplanting (DAT)

| **Models** | **α** | **β** | **γ** | **RMSE**[a] | **$R^2$** | **NMBb** |
|---|---|---|---|---|---|---|
| Calibration | | | | | | |
| Model 1 LNC = **α**SPAD + **β**DAT + **γ** | 1.02 ± 0.07 | -0.31 ± 0.01 | 7.01 ± 2.95 | 3.07 g $kg^{-1}$ | 0.78 | |
| Model 2 LNC = **α**b* + **β**DAT + **γ** | -0.67 ± 0.06 | -0.29 ± 0.01 | 69.27 ± 2.10 | 3.36 g $kg^{-1}$ | 0.71 | |
| Model 3 SPAD = **α**b* + **β**DAT + **γ** | -0.60 ± 0.04 | 0.024 ± 0.007 | 59.66 ± 1.14 | 1.62 | 0.55 | |
| Model 4 SPAD = **α**b* + **γ** | -0.60 ± 0.04 | | 60.70 ± 1.12 | 1.62 | 0.53 | |
| Validation | | | | | | |
| Model 1 | | | | 2.43 g $kg^{-1}$ | 0.75 | 1.19% |
| Model 2 | | | | 2.59 g $kg^{-1}$ | 0.62 | -1.32% |
| Model 3 | | | | 2.01 | 0.46 | -1.94% |
| Model 4 | | | | 1.89 | 0.47 | -2.00% |

The dataset in 2011 was used for model calibration and the dataset in 2010 used for model validation.

[a]Root mean square error.

[b]Normalized mean bias.

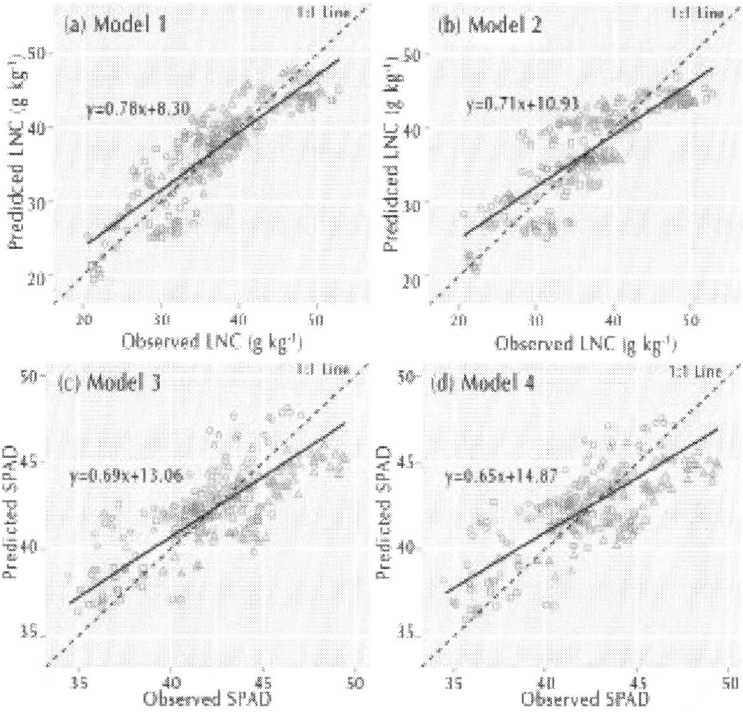


**Figure 4**: Calibration of the four models in Table 4 for the estimation of leaf nitrogen concentration (LNC, g kg-1) and SPAD. Model 1(a), $LNC = \alpha SPAD + \beta DAT + \gamma$, model 2 (b), $LNC = \alpha b^* + \beta DAT + \gamma$, model 3 (c), $SPAD = \alpha b^* + \beta DAT + \gamma$, model 4 (d), $SPAD = \alpha b^* + \gamma$. Different colors denote different developmental stages (red: vegetative, green: tillering, blue: jointing, purple: booting). Different symbols denote different cultivars used for model calibration (○ Liangyoupeijiu, Δ Nanjing45, □ Nanjing46).

Validations were performed on the four models in Table 4 with all data obtained in 2010 under sunny days. The images used in model calibration and validation were taken under different weather conditions, the objective of this combination was to evaluate whether the model was robust under different light conditions. In general, good performances on the predicted models were observed for the estimation of LNC and SPAD (Figure 5). Model 1 showed the best performance on the prediction of LNC with a normalized mean bias of 1.19% (Table 4 and Figure 5a). Model 2, 3 and 4 showed relatively lower $R^2$ and smaller negative bias. The model for the prediction of LNC with color index $b^*$ (Figure 5b) was not severely affected by the different light conditions in 2010 and 2011, compared to the prediction of LNC with SPAD (Figure 5a). As with the calibration results, most of the data in the vegetative stage were below the 1:1 line in models 1 and 2. In addition, the data in tillering stage deviated from the 1:1 line in models 3 and 4, which resulted in the low $R^2$ (Figure 5). The similar $R^2$ and RMSE in Figure 5c and d indicated that the relationship between SPAD and color index *b** was not affected by the developmental stage.

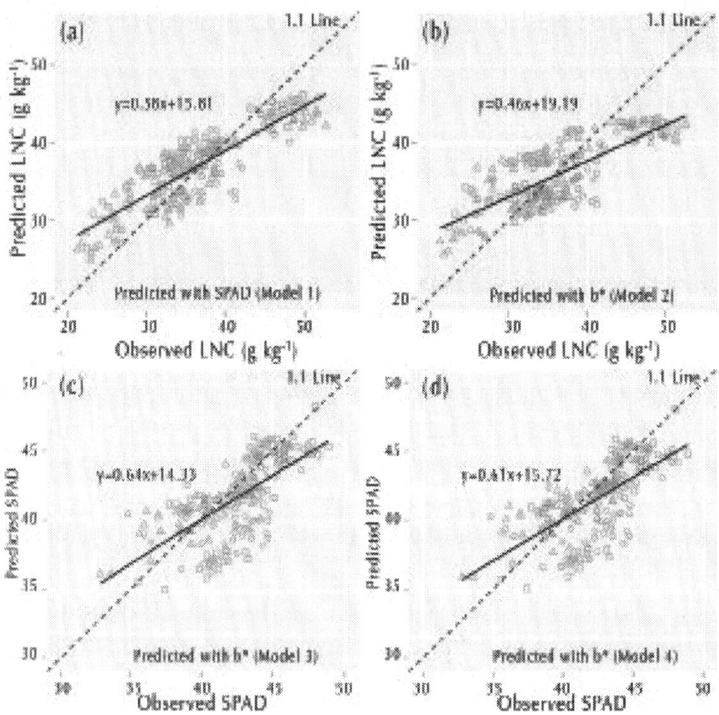


**Figure 5:** Validation of the four models using the fitted parameters in Table-4for the estimation of leaf nitrogen concentration (LNC, g $kg^{-1}$) and SPAD. Model 1 (a), $LNC = \alpha SPAD + \beta DAT + \gamma$, model 2 (b), $LNC = \alpha b^* + \beta DAT + \gamma$, model 3 (c), $SPAD = \alpha b^* + \beta DAT + \gamma$, model 4 (d), $SPAD = \alpha b^* + \gamma$. Different colors denote different developmental stages (red: vegetative, green: tillering, blue: jointing, purple: booting). Different symbols denote different cultivars used for model validation (○ Nanjing44, Δ Wuyunjing24, □ Yangjing48).

In our study, the light conditions during image acquisition were different between 2010 and 2011. Image acquisition in 2011 was under overcast days with low but stable diffused light. In contrast, images were taken under sunny days in 2010 with strong and variable (PAR ranging from 937 to 1759 $\mu$mol m$^{-2}$ s$^{-1}$) light (Table 3). We can remove the effect of illumination change on images when the change is over the whole image by normalizing the image [49] or using a calibration panel [36]. However, there were many other differences caused by the different light conditions. The strong illumination in 2010 caused many white spots and shadows which affected the image color while this phenomenon was hardly seen in the images taken in 2011. With the rice growth, the light status within rice canopy (transmittance, reflectance, absorption) becoming more and more complex, the change of illumination will cause different degrees of influence on canopies that in different height. In this case, it is difficult to calibrate the image color accurately. Sakamoto *et al.*[15] calibrated image indices with the introduction of exposure value (EV) and obtained reliable camera retrieved vegetation indices (VIs). However, this method cannot apply to our study for the calibration of canopy color because of the different influence in one image.

Besides the light conditions, the prediction of crop N status with leaf color may be affected by many other environmental factors, such as developmental stage, diseases and drought stress [50]. These factors may be detectable from high-resolution canopy images [22,23]. Color indices analysis associated with these factors and other image characteristics (e.g. canopy cover, plant shape, leaf texture or even soil status), will provide more reliable results to N diagnosis. Further studies will be devoted to the exploration of image characteristics, leaf color correction and the calibration of the established model with physiological parameters (e.g. SLW) in the evaluation of N status under different environmental conditions.

# CONCLUSIONS

Image color indices calculated from RGB, HSV and $L^*a^*b^*$ color models have significant correlations with SPAD readings and leaf N concentration (LNC) of rice leaves. Among these color indices, the index $b^*$, which represents the visual perception of yellow-blue chroma, had the highest correlation coefficients with SPAD readings and LNC. Regression analysis showed significant linear relationships between index $b^*$ and N parameters. However, the relationship between

LNC and SPAD reading, LNC and index $b^*$ were affected by the rice developmental stage. This is mainly caused by the leaf thickness which can be different in cultivars, developmental stages and environmental conditions. In this case, linear regression models were established between color index $b^*$, LNC and SPAD readings by considering the developmental process in rice. The multiple linear models improved the $R^2$ of SPAD readings and $b^*$ with LNC, yet most of the predicted LNC in the vegetative stage were underestimated because of the inconsistent relationship between chlorophyll and N concentration. Validations on the models showed good performance and acceptable predicted precision with different cultivars and sampling dates under different natural light conditions. These results indicated that digital color image analysis could be a simple method for assessing rice N status under natural light conditions.

# MATERIALS AND METHODS

## General Information of the Experimental Site

The experiment was laid out at Changshu Agricultural Ecology Experiment Station, Changshu, Jiangsu, China (31°33′N, 120°42′E). Located in the humid subtropical climate zone, the station receives average annual solar radiation of 4930 MJ $m^{-2}$, sunshine of 1800 hours, precipitation of 1200 mm and cumulative temperature above 10°C of 4933 degree-days (°C·d). The soil type for the field experimental site is a gleyed paddy soil of the Taihu Lake region, which contains total nitrogen (N) of 1.79 g $kg^{-1}$, total phosphorus (P) of 0.93 g $kg^{-1}$, total potassium (K) of 18.7 g $kg^{-1}$, organic matter of 30.8 g $kg^{-1}$, alkali-extractable N of 123 mg $kg^{-1}$, Olsen-P of 13.1 mg $kg^{-1}$, plant available K of 121 mg $kg^{-1}$ and pH of 7.4 (soil: water, 1:2) in the 0–15 cm soil layer.

## Experimental Design

Two independent experiments with different N fertilization gradients were implemented in our study. Experiment I was a long-term site-specific rice-wheat rotation experiment that started in 1997. The trial comprised six fertilizer treatments represented as CK, N0, N1, N2, N3, and N4 for N application of 0, 0, 180, 225, 270 and 315 kg N $ha^{-1}$ in rice season, respectively. Each treatment had four replicates which were arranged in a randomized block design. The data used in this paper were from the period May to November in 2010 and 2011 with cultivars Wuyunjing24 and Nanjing46, respectively. Experiment II was carried out in paddy fields with a rice-wheat rotation in 2010 and 2011. Six N application rates with three replicates were designed in this trial, which were represented as N0, N1, N2, N3, N4 and N5 with N application of 0, 120, 180, 240, 270 and 300 kg N $ha^{-1}$, respectively. The cultivars were Nanjing44 and Yangjing48 in 2010, and Nanjing45 and Liangyoupeijiu in 2011. For both experiments, the N was split into three applications, 40% as basal, 20% at tillering and 40% at booting. In addition, each plot received 90 kg K $ha^{-1}$ and 20 kg P $ha^{-1}$ except the CK treatment in experiment I. The applied K was split into 50% as basal and 50% at booting, and all the P was applied as basal fertilizer. Other crop managements were same as the local traditional practices.

## Sample Collection and Digital Image Acquisition

For measuring rice growth and nutrition parameters, the above-ground part of rice plant was sampled about every two weeks after transplanting until the booting stage. A total of 4 sets of samples were collected in 2010 and 2011 (Table 3). The plant samples were separated into leaves and stems (including sheaths), and dried at 105°C for half an hour and then at 70°C until constant weight. After that, the samples were weighed for dry weight and analyzed for leaf N concentration (LNC) by the Kjeldahl method [51]. Along with the plant sampling, a chlorophyll meter (SPAD-502, Minolta Camera Co., Osaka, Japan) was used to obtain SPAD values on the four youngest fully expanded leaves. Each blade was measured at three points: on the upper, middle and lower thirds on either side of the midrib. Then, average SPAD readings were calculated for each plot.

On the same day or following day of plant sampling, images of the rice canopy were captured using a digital still color camera (EOS 50D, Canon Inc.) with a resolution of 15 mega pixels. The camera was mounted on a tripod at the nadir position with a constant height of 1 m above the top of the rice canopy. Aperture priority mode was selected, and the camera was set at aperture of f/5.6, ISO of 100, white balance of 4,900 K, auto exposure and auto-focus with the flash turned off. In 2010, the pictures were taken at local time 12:00 – 13:00 in sunny days, while in 2011, the pictures were taken at the same time period but on overcast days. In the days of picture taken (July and August), the deviation between local time and solar noon was within 4 minutes. All the pictures from the experiments were stored in CR2 (Canon raw image file) format. The photosynthetic active radiation (PAR) and illuminance were recorded by a portable light meter (GLZ-C, Top Instrument Co., Zhejiang, P. R. China) during the period of image acquisition. Average PAR was calculated with each set of pictures, and observed 937–1759 $\mu$mol photons $m^{-2}$ $s^{-1}$ and 145–692 $\mu$mol photons $m^{-2}$ $s^{-1}$ in 2010 and 2011, respectively (Table 3).

## Image Segmentation and Color Indices Calculation

A raw image file contains minimally processed data from the image sensor of a digital camera. This file saves settings of white balance, color saturation, contrast and sharpness in it, but defers the processing. Therefore, all the modification made on a raw image file is non-destructive.

The canopy images in CR2 format were adjusted for white balance using the 18% gray card (R-27, Kodak) pictures which were taken simultaneously with the canopy images. Then, lens distortion correction was applied, and exposure was set to +1 for all images. After that, images were saved as joint photographic experts group (JPEG) files for further processing. All the procedures above were processed with Adobe Camera Raw (Adobe Systems Inc.).

Since the images contained the rice canopy and some non-canopy elements, such as soil, water and plant residues, images were segmented into canopy portion and non-canopy portion. A computer program was developed based on the G-R thresholding method [25,29] using MatLab® (MathWorks Inc.) to extract the canopy portion of the image. The G-R thresholding method was proposed according to the difference of reflectance spectrum between green vegetation and non-canopy elements in the visible band. There is a reflection peak for green vegetation in the green band, whereas no apparent change for soil or water albedo in the whole visible band. Therefore, the value of green channel minus that of red channel expands the difference between canopy and non-canopy portion.

After the image segmentation, 13 color indices derived from 3 color models were calculated. RGB model is the most common color model for the representation of digital images. A color in the RGB model is described by indicating how much of each of the red, green, and blue is included. The color is expressed as an RGB triplet (R, G, B), with the representation for black of (0, 0, 0) and for the brightest representable white of (255, 255, 255) in an 8-bit image [52]. R, G and B are the mean values of the red, green and blue channels, and r, g and b are the normalized RGB values, respectively. Intensity (INT) is the average of R, G and B. $VI_{Green}$ is a widely used vegetation index [13]. These indices were calculated as follows [27,29]:

$$r=R/(R+G+B) \tag{1}$$

$$g=G/(R+G+B) \tag{2}$$

$$b=B/(R+G+B) \tag{3}$$

$$INT=(R+G+B)/3 \tag{4}$$

$$VI_{Green}=(G-R)/(G+R) \tag{5}$$

In addition, the CIE $L^*a^*b^*$ and HSV color spaces were also tested in this study. The $L^*$ coordinate in CIE $L^*a^*b^*$[53] closely matches human perception of lightness, $a^*$ and $b^*$ dimensions represent the visual perception of red-green and yellow-blue chroma, respectively. Both $a^*$ and $b^*$ are independent with image lightness ($L^*$), and take on both negative and positive values ($+a^*$ reds, $-a^*$ greens, $+b^*$ yellows, $-b^*$ blues). The three coordinates of $L^*a^*b^*$ are computed from the tristimulus values X, Y and Z as following equations [32, 54]:

$$L^*=116f(Y/Y_n)-16 \quad (6)$$

$$a^*=500[f(X/X_n)-f(Y/Y_n)] \quad (7)$$

$$b^*=200[f(Y/Y_n)-f(Z/Z_n)] \quad (8)$$

$$f(\omega)=\begin{cases}(\omega^{1/3}) & |\omega<0.008856 \\ 7.787(\omega)+16/116 & |\omega\geq 0.008856\end{cases} \quad (9)$$

where $X_n$, $Y_n$ and $Z_n$ describe a specified white object-color stimulus.

The HSV color space is represented as a cylindrical-coordinate in which the angle around the central vertical axis corresponds to hue (H). The calculation of H was listed below [27,29]:

$$H=\begin{cases}60*\{(G-B)/[\max(RGB)-\min(RGB)]\}, \max(RGB)=R \\ 60*\{2+\{(B-R)/[\max(RGB)-\min(RGB)]\}\}, \max(RGB)=G \\ 60*\{4+\{(R-G)/[\max(RGB)-\min(RGB)]\}\}, \max(RGB)=B\end{cases} \quad (10)$$

Pearson correlation and regression analyses were used to detect the relationship between color indices and crop N status. The significance of linear regressions was evaluated using Student's *t*-test at 95% confidence levels. Significance of ANOVAs was evaluated with the least significant difference test (LSD) at 0.05 probability level. Data analysis and figure production were done using the R v3.0.3 software [55].

Correlation analysis, linear regression analysis and model establishment between color indices and crop N status were based on the data in 2011, and the data in 2010 were used for model validation.

# ACKNOWLEDGEMENTS

This research was supported by the National Key Technologies R & D Program of China (2012BAJ24B06). The authors acknowledge John Angus of CSIRO Agriculture for his constructive comments and revisions of the manuscript. We wish to thank the editor and two anonymous reviewers for their very helpful comments on earlier drafts.

# REFERENCES

1. Fageria NK, Baligar VC, Jones CA: *Growth and Mineral Nutrition of Field Crops*. 3rd edition. Boca Raton: CRC Press; 2011.
2. Adhikari C, Bronson KF, Panuallah GM, Regmi AP, Saha PK, Dobermann A, Olk DC, Hobbs PR, Pasuquin E: On-farm soil N supply and N nutrition in the rice–wheat system of Nepal and Bangladesh. *Field Crops Res* 1999, 64:273-286.
3. Prasertsak A, Fukai S: Nitrogen availability and water stress interaction on rice growth and yield. *Field Crops Res* 1997, 52:249-260.
4. Kaushal SS, Groffman PM, Band LE, Elliott EM, Shields CA, Kendall C: Tracking nonpoint source nitrogen pollution in human-impacted watersheds. *Environ Sci Technol* 2011, 45:8225-8232.
5. Zhang F, Cui Z, Fan M, Zhang W, Chen X, Jiang R: Integrated soil–crop system management: reducing environmental risk while increasing crop productivity and improving nutrient use efficiency in China. *J Environ Qual* 2011, 40:1051-1057.
6. Miao YX, Stewart BA, Zhang FS: Long-term experiments for sustainable nutrient management in China: a review. *Agron Sustain Dev* 2011, 31:397-414.
7. Li F, Mistele B, Hu Y, Chen X, Schmidhalter U: Reflectance estimation of canopy nitrogen content in winter wheat using optimised hyperspectral spectral indices and partial least squares regression. *Eur J Agron* 2014, 52:198-209.
8. Chen P, Haboudane D, Tremblay N, Wang J, Vigneault P, Li B: New spectral indicator assessing the efficiency of crop nitrogen treatment in corn and wheat. *Remote Sens Environ* 2010, 114:1987-1997.
9. Peng S, Buresh RJ, Huang J, Zhong X, Zou Y, Yang J, Wang G, Liu Y, Hu R, Tang Q, Cui K, Zhang F, Dobermann A: Improving nitrogen fertilization in rice by site-specific N management: a review. *Agron Sustain Dev* 2010, 30:649-656.

10. Lin FF, Qiu LF, Deng JS, Shi YY, Chen LS, Wang K: Investigation of SPAD meter-based indices for estimating rice nitrogen status. *Comput Electron Agric* 2010, 71(Supplement 1):S60-S65.
11. Cabangon RJ, Castillo EG, Tuong TP: Chlorophyll meter-based nitrogen management of rice grown under alternate wetting and drying irrigation. *Field Crops Res* 2011, 121:136-146.
12. Reyniers M, Walvoort DJJ, De Baardemaaker J: A linear model to predict with a multi-spectral radiometer the amount of nitrogen in winter wheat. *Int J Remote Sens* 2006, 27:4159-4179.
13. Li Y, Chen D, Walker CN, Angus JF: Estimating the nitrogen status of crops using a digital camera. *Field Crops Res* 2010, 118:221-227. Publisher Full Text
14. Sakamoto T, Shibayama M, Kimura A, Takada E: Assessment of digital camera-derived vegetation indices in quantitative monitoring of seasonal rice growth. *ISPRS J Photogramm Remote Sens* 2011, 66:872-882.
15. Sakamoto T, Gitelson AA, Nguy-Robertson AL, Arkebauer TJ, Wardlow BD, Suyker AE, Verma SB, Shibayama M: An alternative method using digital cameras for continuous monitoring of crop status. *Agric For Meteorol* 2012, 154–155:113-126.
16. Blackmer TM, Schepers JS: Use of a chlorophyll meter to monitor nitrogen status and schedule fertigation for corn. *J Prod Agric* 1995, 8:56-60.
17. Rorie RL, Purcell LC, Mozaffari M, Karcher DE, King CA, Marsh MC, Longer DE: Association of "greenness" in corn with yield and leaf nitrogen concentration. *Agron J* 2011, 103:529-535.
18. Zhang J, Blackmer AM, Ellsworth JW, Koehler KJ: Sensitivity of chlorophyll meters for diagnosing nitrogen deficiencies of corn in production agriculture. *Agron J* 2008, 100:543-550.
19. Zhang C, Kovacs JM: The application of small unmanned aerial systems for precision agriculture: a review. *Precis Agric* 2012, 13:693-712.
20. Samborski SM, Tremblay N, Fallon E: Strategies to make use of plant sensors-based diagnostic information for nitrogen recommendations. *Agron J* 2009, 101:800-816.
21. Scharf PC, Lory JA: Calibrating corn color from aerial photographs to predict sidedress nitrogen need. *Agron J* 2002, 94:397-404.

22. Berger B, Parent B, Tester M: High-throughput shoot imaging to study drought responses. *J Exp Bot* 2010, 61:3519-3528.

23. Fanourakis D, Briese C, Max JF, Kleinen S, Putz A, Fiorani F, Ulbrich A, Schurr U: Rapid determination of leaf area and plant height by using light curtain arrays in four species with contrasting shoot architecture. *Plant Methods* 2014, 10:9.

24. Golzarian MR, Frick RA, Rajendran K, Berger B, Roy S, Tester M, Lun DS: Accurate inference of shoot biomass from high-throughput images of cereal plants. *Plant Methods* 2011, 7:1-11.

25. Wang Y, Wang D, Zhang G, Wang C: Digital camera-based image segmentation of rice canopy and diagnosis of nitrogen nutrition. *Trans Chin Soc Agric Eng* 2012, 28:131-136.

26. Abramoff MD, Magalhães PJ, Ram SJ: Image processing with Image J. *Biophotonics Int* 2004, 11:36-42.

27. Lee K-J, Lee B-W: Estimation of rice growth and nitrogen nutrition status using color digital camera image analysis. *Eur J Agron* 2013, 48:57-65.

28. Liu J, Pattey E: Retrieval of leaf area index from top-of-canopy digital photography over agricultural crops. *Agric For Meteorol* 2010, 150:1485-1490.

29. Wang Y, Wang D, Zhang G, Wang J: Estimating nitrogen status of rice using the image segmentation of G-R thresholding method. *Field Crops Res* 2013, 149:33-39.

30. Hunt ER Jr, Doraiswamy PC, McMurtrey JE, Daughtry CS, Perry EM, Akhmedov B: A visible band index for remote sensing leaf chlorophyll content at the canopy scale. *Int J Appl Earth Obs Geoinf* 2013, 21:103-112.

31. Pagola M, Ortiz R, Irigoyen I, Bustince H, Barrenechea E, Aparicio-Tejo P, Lamsfus C, Lasa B: New method to assess barley nitrogen nutrition status based on image colour analysis: comparison with SPAD-502. *Comput Electron Agric* 2009, 65:213-218.

32. Wiwart M, Fordoński G, Żuk-Gołaszewska K, Suchowilska E: Early diagnostics of macronutrient deficiencies in three legume species by color image analysis. *Comput Electron Agric* 2009, 65:125-132.

33. Vollmann J, Walter H, Sato T, Schweiger P: Digital image analysis and chlorophyll metering for phenotyping the effects of nodulation in soybean. *Comput Electron Agric* 2011, 75:190-195.

34. Golzarian MR, Frick RA: Classification of images of wheat, ryegrass and brome grass species at early growth stages using principal component analysis. *Plant Methods* 2011, 7:1-11.

35. McCarthy CL, Hancock NH, Raine SR: Applied machine vision of plants: a review with implications for field deployment in automated farming operations. *Intell Serv Rob* 2010, 3:209-217.

36. Noh H, Zhang Q, Han S, Shin B, Reum D: Dynamic calibration and image segmentation methods for multispectral imaging crop nitrogen deficiency sensors. *Trans ASAE* 2005, 48:393-401.

37. Afshari-Jouybari H, Farahnaky A: Evaluation of Photoshop software potential for food colorimetry. *J Food Eng* 2011, 106:170-175.

38. Graeff S, Pfenning J, Claupein W, Liebig HP: Evaluation of image analysis to determine the N-fertilizer demand of broccoli plants (Brassica oleracea convar. botrytis var. italica). *Adv Opt Technol* 2008, 2008:1-8.

39. Xue L, Cao W, Luo W, Dai T, Zhu Y: Monitoring leaf nitrogen status in rice with canopy spectral reflectance. *Agron J* 2004, 96:135-142.

40. Huang J, He F, Cui K, Buresh RJ, Xu B, Gong W, Peng S: Determination of optimal nitrogen rate for rice varieties using a chlorophyll meter. *Field Crops Res* 2008, 105:70-80.

41. Zong S, Lu C, Zhao L, Wang C, Dai Q, Zou J: Physiological basis of high yield of an inter-subspecific hybrid rice, Liangyoupeijiu. *J Nanjing Agric Tech Coll* 2000, 16:8-12.

42. Shibayama M, Watanabe Y: Estimating the mean leaf inclination angle of wheat canopies using reflected polarized light. *Plant Prod Sci* 2007, 10:329-342.

43. Homem Antunes MA, Walter-Shea EA, Mesarch MA: Test of an extended mathematical approach to calculate maize leaf area index and leaf angle distribution. *Agric For Meteorol* 2001, 108:45-53.

44. Chapman SC, Barreto HJ: Using a chlorophyll meter to estimate specific leaf nitrogen of tropical maize during vegetative growth. *Agron J* 1997, 89:557-562.

45. Yang WH, Peng S, Huang J, Sanico AL, Buresh RJ, Witt C: Using leaf color charts to estimate leaf nitrogen status of rice. *Agron J* 2003, 95:212-217.

46. Peng S, García FV, Laza RC, Cassman KG: Adjustment for specific leaf weight improves chlorophyll meter's estimate of rice leaf nitrogen concentration. *Agron J* 1993, 85:987-990.

47. Shukla AK, Ladha JK, Singh V, Dwivedi B, Balasubramanian V, Gupta RK, Sharma S, Singh Y, Pathak H, Pandey P: Calibrating the leaf color chart for nitrogen management in different genotypes of rice and wheat in a systems perspective. *Agron J* 2004, 96:1606-1621.

48. Baret F, Fourty T: Estimation of leaf water content and specific leaf weight from reflectance and transmittance measurements. *Agronomie* 1997, 17:455-464.

49. Drew MS, Wei J, Li ZN: Illumination–invariant image retrieval and video segmentation. *Pattern Recognit* 1999, 32:1369-1388.

50. Pydipati R, Burks TF, Lee WS: Identification of citrus disease using color texture features and discriminant analysis. *Comput Electron Agric* 2006, 52:49-59.

51. Lu RK: *Soil Chemical Nutrient Analysis Techniques*. Beijing: China Agriculture and Technology Press; 2000.

52. Solomon C, Breckon T: *Fundamentals of Digital Image Processing: a Practical Approach with Examples in Matlab*. Chichester: John Wiley & Sons; 2011.

53. Robertson AR: The CIE 1976 color-difference formulae. *Color Res Appl* 1977, 2:7-11.

54. Fairchild MD: *Color Appearance Models*. 2nd edition. Chichester: John Wiley & Sons; 2005.

55. R Core Team: *R: A language and environment for statistical computing*. Vienna, Austria: R Foundation for Statistical Computing; 2014.

Chapter 5

# Callosemeasurer: A Novel Software Solution to Measure Callose Deposition and Recognise Spreading Callose Patterns

Ji Zhou[1], Thomas Spallek[1,2], Christine Faulkner[1,3] and Silke Robatzek[1*]

[1]The Sainsbury Laboratory, Norwich Research Park, Norwich, NR4 7UH, UK

[2]Present address: RIKEN Yokohama Institute, Suehiro-cho, Tsurumi-ku, Yokohama City, Kanagawa, 230-0045, Japan

[3]Present address: Department of Biological and Medical Sciences, Oxford Brookes University, Oxford, OX3 0BP, UK

# ABSTRACT

## Background

Quantification of callose deposits is a useful measure for the activities of plant immunity and pathogen growth by fluorescence imaging. For robust scoring of differences, this normally requires many technical and biological replicates and manual or automated quantification of the callose deposits. However, previously available software tools for quantifying callose deposits from bioimages were limited, making batch processing of callose image data problematic. In particular, it is challenging to perform large-scale analysis on images with high background noise and fused callose deposition signals.

## Results

We developed CalloseMeasurer, an easy-to-use application that quantifies callose deposition, a plant immune response triggered by potentially pathogenic microbes. Additionally, by tracking identified callose deposits between multiple images, the software can recognise patterns of how a given filamentous pathogen grows in plant leaves. The software has been evaluated with typical noisy experimental images and can be automatically executed without the need for user intervention. The automated analysis is achieved by using standard image analysis functions such as image enhancement, adaptive thresholding, and object segmentation, supplemented by several novel methods which filter background noise, split fused signals, perform edge-based detection, and construct networks and skeletons for extracting pathogen growth patterns. To efficiently batch process callose images, we implemented the algorithm in C/C++ within the Acapella™ framework. Using the tool we can robustly score significant differences between different plant genotypes when activating the immune response. We also provide examples for measuring the *in planta* hyphal growth of filamentous pathogens.

## Conclusions

CalloseMeasurer is a new software solution for batch-processing large image data sets to quantify callose deposition in plants. We demonstrate its high accuracy and usefulness for two applications: 1) the quantification of callose deposition in different genotypes as a measure for the activity of plant immunity; and 2) the quantification and detection of spreading networks of callose deposition triggered by filamentous pathogens as a measure for growing pathogen hyphae. The software is an easy-to-use protocol which is executed within the Acapella software system without requiring any additional libraries.

# BACKGROUND

Immunity against potentially infectious pathogens in plants involves a plethora of defence responses such as the deposition of callose, a 1–3 β-linked glucan polymer [1,2]. Imaging callose deposition has emerged as a widely used method to quantify the activity of plant defences to a range of different pathogens and pathogen-derived molecules (e.g. flg22 derived from bacterial flagellin) in different plant genotypes and mutants [3,4]. Measuring callose deposition is also a popular way to determine the activity of pathogen-derived virulence factors that interfere with the plant immune pathways to the benefit of the pathogen [5,6]. While the principle method of callose staining with aniline blue, followed by clearance of the leaves and taking microscopy images under UV light is well established [7], this approach is hampered by the fact that callose deposits can differ between replicate samples due to biological variation. Moreover, the pattern of spreading callose deposits can vary in response to different pathogen species as well as modes of infection [8,9]. To take these variations and differences into consideration, it is necessary to acquire a larger number of images, use more accurate solutions to quantify callose deposition and to measure pathogen growth patterns.

Improved quantification methods based on automated large-scale image processing will provide better measurements of defence responses, allowing the detection of subtle differences and thereby promoting our understanding of the mechanisms of plant immunity. The usefulness of quantitative bioimage analysis has been demonstrated for high-throughput microscopy in plant endomembrane trafficking [10,11] and monitoring plasmodesmata development [12]. Software solutions developed in these studies allowed comparative measurements of endosomal compartments and plasmodesmata revealing significant differences between different plant genotypes, chemical treatments, biotic and abiotic stresses and during plant development that in many cases were not possible to be observed by the human eye [10,12].

To date, measurements of callose deposits mostly rely on ImageJ [13] and FIJI [14] and/or some related plugins to extract quantifiable data from images of aniline-blue stained leaves [15,16]. Another emerging software package that contains similar functions for quantifying particle-like objects is ICY [17]. Although these software tools enable the detection of callose signals from microscope images, they are limited in their ability to accurately measure callose deposits. For example, we utilised both FIJI and ICY to process a typical callose image. We followed the image processing workflow previously published and applied "Auto Threshold" and "Particle Analyze" functions in FIJI and the "Spots Detector" method in ICY. The results suggest that the two software tools lacked sufficient functions to filter false detected objects as well as to reliably conduct shape/size measurements on detected callose deposits. The results were even more erroneous whilst batch processing callose images (e.g. using macro scripting in FIJI and selecting the "batch input detection" mode in ICY). Because most plant microscopy images contain autofluorescing noise signals derived from chloroplasts, xylem vessels, trichomes, and/or out of focus particle-like signals (typical background signals for plant leaf images), the current available image analysis tools can lead to incorrect detection and imprecise size/shape measurements. Furthermore, in practice these software tools are still semi-automated – manual inputs are required to enhance image quality, choose thresholding algorithms, and/or adjust filtering methods, which makes the image processing of callose deposition time consuming, error-prone, and not applicable for batch processing.

To overcome the above limitations, we developed CalloseMeasurer (v1.0) – a robust software solution that can automate the detection of callose deposits with a very high degree of accuracy and also recognise growth patterns of filamentous pathogen species. This software is based on the Acapella software framework (V2.0, PerkinElmer), which is designed for performing high content and high-throughput bioimage analysis. The usefulness and applicability of CalloseMeasurer are demonstrated with two example experiments.

# RESULTS AND DISCUSSION

## Algorithm

Similar to other image analysis approaches, we applied spot detection algorithms [18] to identify particle-like signals. To robustly filter background noise, we chose intensity analysis and adaptive thresholding methods [19] for removing fluorescent noise. To conduct sensitive edge-based measurements on size and shape, we implemented an edge detection method based on the border tracing approach [20]. Lastly, with the aim of recognising patterns of spreading callose deposits (i.e. filamentous pathogen growth patterns), we designed a tailored function to construct networks for spreading callose and extract skeletons of these networks, which was derived from the *network snakes* approach [21]. We integrated these methods into a powerful software solution that can filter out noise signals, split fused fluorescence signals, measure the size/shape of identified callose objects, and recognise patterns of spreading callose. In practice, we embedded the solution in a workflow for batch processing pathogen-induced callose images, which includes three main phases: detecting regions of interest (ROI) (Figure 1), measuring callose deposits (Figure 2), and recognising patterns of spreading callose deposits (Figure 3). Quantifiable results generated by CalloseMeasurer are saved in two CSV files (one containing results for every processed image and one for overall results).

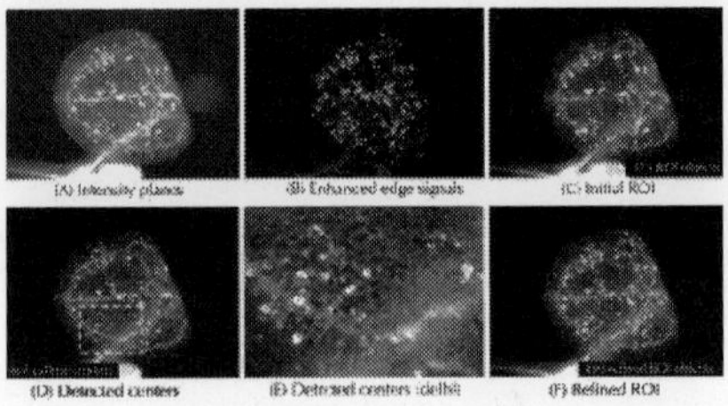


**Figure 1:** The CalloseMeasurer analysis workflow for recognising ROI objects. (A) The algorithm reads a series of callose image files into the software system, which are split into three planes – hue, saturation, and intensity value. Only intensity planes are used in the analysis. (B) As leaf vessel and mesophyll cell signals contain high intensity values, in order to differentiate them with callose deposition signals, images are transformed into their gradients so that callose edges can be highlighted. (C) Based on the processed images, image masks are applied to recognise ROI, which are randomly coloured. (D, E) Within the ROI, the algorithm detects centres of callose through finding local maxima of intensity. (F) ROI objects are rebuilt based on the detected callose centres.

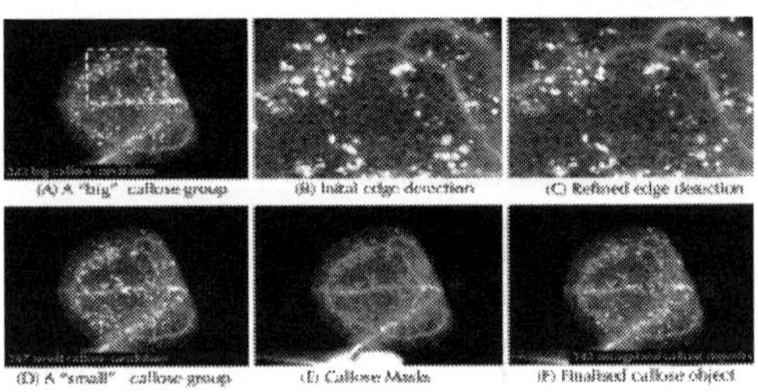


**Figure 2:** The CalloseMeasurer analysis workflow for measuring callose deposits. (A) The reconstructed ROI objects (Figure 2F) are divided into two groups – a "big" callose group (A) and a "small" callose group (D), both are randomly coloured. (B, C) In the "big" callose group, the border of callose is recalculated (callose objects are merged/separated based on the callose centres shown in Figure 2D). Some wrongly recognised callose candidates are removed. (D) Similar to the "big" callose group, the border of every small callose is measured and some wrongly detected ones are removed. (E, F) The refined "small" and "big" callose groups are merged (coloured red) and treated as callose deposition objects (randomly coloured). Quantifiable results (e.g., size, circularity, and intensity) are exported to two CSV files – one for every processed image and one for overall results.

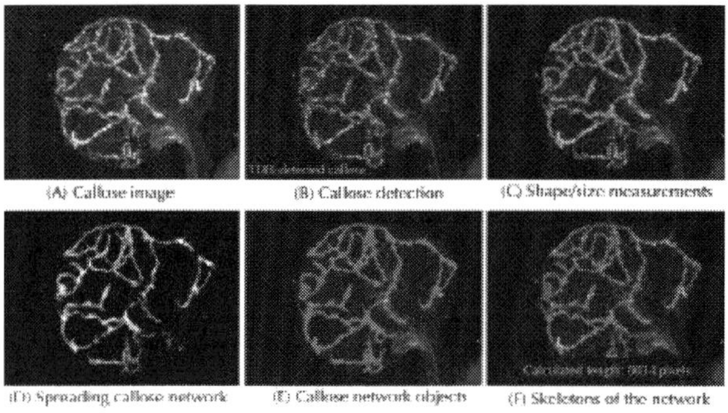


**Figure 3:** The CalloseMeasurer analysis workflow for detecting callose spreading networks. (A, B) If a user ticks the "Detect Callose Network" selection, a series of image files are read into the system and callose deposits (randomly coloured) will be detected by the algorithm. In this case, both small and big callose deposits are identified and used for constructing spreading callose networks. (C) Size, circularity, and intensity of the recognised callose deposits (coloured green) are measured. (D) A spreading callose network is constructed based on the recognised callose deposits. (E) Image masks are applied to the network (coloured red). (F) A filtering system is used to screen out unsuitable masks based on size, width, length, and shape. Skeletons are extracted from the refined callose network. Analysis results (e.g., the size and spreading length of callose deposits) are calculated and exported to the CSV file, which contains overall analysis results.

## Implementation

We implemented the CalloseMeasurer algorithm in C/C++ together with a number of basic image analysis functions provided by the Acapella image library. In order to detect ROI, the algorithm reads callose images into the Acapella system and then divides them into three planes (e.g. hue, saturation, and intensity value planes). Only intensity planes are used during the image analysis (Figure 1A). As most of the input images contain high leaf vessel and mesophyll cell signals, the intensity planes are transformed into their gradients so that the border of callose signals can be highlighted (Figure 1B). Based on the processed images, a watershed method and image masks are applied to identify ROI (coloured randomly in Figure 1C), within which centres of every ROI object are located through the detection of local maxima of intensity (see Figure 1D and 1E). Centres with low intensity/contrast values are removed and remaining ones are treated as centres of callose deposit candidates. By taking into account raw image data (i.e. the intensity planes), recognised ROI objects, and callose deposition centres, the algorithm splits and rebuilds ROI (Figure 1F). The number of refined ROI objects could be different from the number of callose centres during the reconstruction of the objects list.

As the size and shape of refined ROI objects are precisely measured, in order to perform sensitive measurements on callose deposits, the algorithm firstly divides the ROI objects into two size groups – a "big" callose group (Figure 2A) and a "small" callose group (Figure 2D), both are randomly coloured. For the "big" callose group, the border of every "big" object is recalculated based on their centres and some incorrectly separated (or merged) objects are remerged (or split). In the meantime, objects with low intensity/contrast or odd shapes are removed (Figure 2B and 2C). For the "small" callose group, a similar procedure is followed, which recalculates and filters small callose objects. After this step, refined "big" or "small" objects are merged (coloured red in Figure 2E) and treated as finalised callose deposits (randomly coloured in Figure 2F). Lastly, object measurements are conducted to quantify features such as size (in pixel$^2$), perimeter (in pixel), circularity (according to the size and perimeter of recognised callose objects), and fluorescent signal intensity (between 0 and 255, which are extracted from the original intensity planes). Results are exported to two CSV files during the batch processing – one containing results for every processed image (including processed image name, callose index, size, circularity, and signal intensity) and one for overall results (including processed image name, callose number per image, average size and signal intensity of detected callose deposits).

When implementing the algorithm, we followed best practices in image processing and treated bioimages as data, so that feature selection could be mainly performed based on statistical analysis of features of image data sets [22]. As software filters (e.g. convolution and median filters) degrade biological image data, therefore we only extracted quantifiable results (e.g. intensity/contrast) from raw image data. In order to differentiate high background signals from callose deposits, we utilised global values at the image level (for processing images) and adaptive local values at the object level (for analysing objects). Moreover, we also applied some software engineering concepts (e.g. reusability and modularity) to the development of the software for improving software usability and functionality.

## Examples of Applications

### *Measuring Callose Deposition in Arabidopsis Thaliana Triggered By Bacterial Flagellin*

The flg22 peptide derived from bacterial flagellin is a powerful agent to trigger callose deposits in plants [23]. We treated Arabidopsis cotyledons for 24 hrs with 1 μM flg22 and imaged fluorescence of aniline blue stained callose deposits [24]. To illustrate the high degree of accuracy of quantifying callose deposits using CalloseMeasurer we compared the flg22 responses from wild type plants and mutants in *flagellin sensing 2* (*fls2*), which encodes the receptor for flg22 and thereby causes insensitivity to this microbe-derived molecule [25]. The algorithm was able to detect 0–473 callose deposits per image sample and measure significant differences (student's t-test) between these genotypes following flg22 treatment (Figure 4).

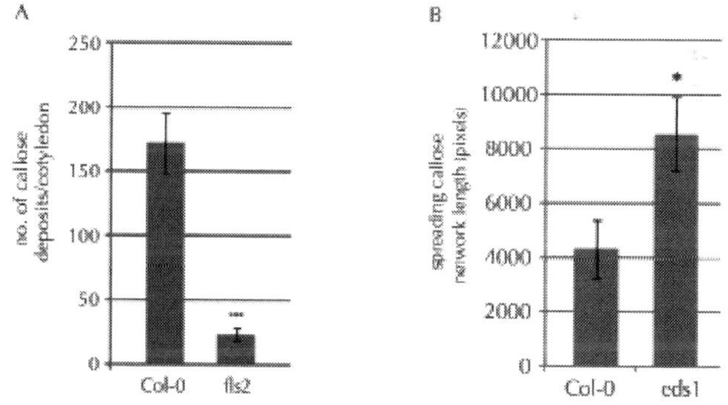


**Figure 4:** Measurement of callose deposition following flg22 treatment and after. *Hpa* infection using CalloseMeasurer. (A) The number of callose deposits following flg22 treatment was significantly greater on Arabidopsis Col-0 cotyledons than on cotyledons of *fls2* mutant plants. (B) The spreading network of callose on *Hpa* infected leaves was significantly longer on *eds1* mutant leaves than on Arabidopsis Col-0 leaves. Asterisks indicate p-values < 0.05 (*) and < 0.001 (***). Error bars represent the standard error.

## Measuring Callose Deposition in Arabidopsis Thaliana Infected with an Oomycete Pathogen

Another strong infection stress that induces the accumulation of callose is the encasement of the haustoria of filamentous pathogens such as the oomycete *Hyaloperonospora arabidopsidis* (*Hpa*) on *Arabidopsis thaliana*[8]. For infecting plants, we spore inoculated seedling leaves with the *Hpa*strain Waco9 as previously described [26] and imaged leaves 6 days post infection. In order to detect the accumulated callose structure, we included a unique function in CalloseMeasurer, detecting "Callose Network", to measure the spreading length of callose deposits. This functions as a biological measure of the total length of the hyphae that has undergone encasement. If the "Callose Network" option is ticked by users, the algorithm will process input images as previously described – detecting and measuring callose deposits from microscope images (Figure 3A-3C). After that, centres of every detected callose deposit are located and relevant centres are connected if the distance between two centres is shorter or equal to a defined distance threshold. In general, users can enter a specific value (in pixels) as the distance threshold. If no value has been entered, CalloseMeasurer will calculate a default distance value, based on which the spreading callose network will be constructed. The formula that calculates the default distance value can be seen as follows:

$$\text{Dis tan ce} = \frac{1}{n}\sum_{i=1}^{n} D_i + \frac{1}{m}\sum_{j=1}^{m} d_j + \frac{1}{l}\sum_{k=1}^{l} F_k \qquad (1)$$

(D*i* is the diameter of the i$^{th}$ object of the "big" callose objects list and n is the size of the list; d*j* is the diameter of the j$^{th}$ object of the "small" callose objects list and m is the size of the list; F*k* is the full length of the k$^{th}$ object of the detected callose objects list and l is the size of the list).

After connecting relevant callose centres, a spreading callose network is created (Figure 3D). Image masks can be applied to the network and CalloseMeasurer will discard masks with unsuitable size, width, length, and shape (Figure 3E). Remaining masks are saved in one objects list so that skeletons of these objects can be extracted (Figure 3F). The software measures the spreading length of callose deposits based on the extracted skeletons. The formula to perform the calculation is:

$$\text{Length} = \sum_{i=1}^{n}\left(\frac{p_i - w_i}{2}\right) \quad (2)$$

(*pi* is the measured perimeter of the i$^{th}$ object of the callose skeleton objects list, *wi* is the measured width of the callose skeleton objects, and n is the size of the callose skeleton objects list).

Furthermore, we quantified the length of hyphae with encased haustoria for Arabidopsis Col-0 wild type and the mutant *enhanced disease susceptibility 1* (*eds1*) (Figure 4). Mutant *eds1* plants are more susceptible to *Hpa* Waco9 when spore counts are measured [27]. Similar results were shown by CalloseMeasurer which detected significantly longer lengths of encased haustoria, indicating a more advanced infection.

## Limitations

Automated detection of callose deposition promises to be useful in many plant-pathogen interaction studies. It relies on a mixture of image analysis tasks such as image enhancement, signal filtering, object segmentation, object measurements, and feature selection. CalloseMeasurer offers the functionality for performing these tasks and allows researchers to robustly and accurately detect callose deposits for monitoring activities of plant immunity. Although the software framework (the Acapella software framework) used in the implementation is a commercial platform, it has been regularly used in bioimage analysis in many research areas [11,28,29]. Moreover, with the aim of sharing the software solution with the research community, we only chose basic image analysis modules/functions from the Acapella library during the development of the software. All functions used have counterparts in open-source bioimage analysis libraries such as ImgLib [30], which means that if necessary, the source code can be easily translated into an open-source software package with limited development efforts. In order to help users or developers to gain an in-depth understanding of CalloseMeasurer, we provide detailed user manual, detailed comments. For testing purpose, users can obtain a free one-month trial version of the Acapella system from PerkinElmer.

# CONCLUSIONS

Along with the increasing importance of bioimage informatics in recent years, many bioinformaticians and computational biologists are dedicating themselves to developing novel computational techniques to extract meaningful data from large-scale biological image data[31,32]. A number of computational techniques such as signal processing, machine learning, data mining, mathematical modelling, and multi-dimensional data visualisation have been utilised to solve various data-intensive biological problems [33]. Following this trend, we previously developed algorithms for quantification of plant cells, plasmodesmata and endomembrane compartments and have now designed and implemented a novel software solution to measure the activity of plant immunity at the tissue level. In this study, we implement CalloseMeasurer, a new software solution for accurate quantification of callose deposits from large sets of bioimages. The unique features of CalloseMeasurer are: batch-processing of images; robust filtering of background noise signals common to plant fluorescent microscopy images; detection and measurement of callose deposits with high sensitivity and accuracy; and detection of spreading networks of callose. For demonstrating the usefulness of CalloseMeasurer, we presented two example applications that show quantitative differences in callose deposition between genotypes and detection of pathogen growth *in planta*.

# METHODS

## Plant Growth Conditions

*Arabidopsis thaliana* plants were grown on Jiffy pellets (Jiffy Products International AS, Norway), or for sterile conditions on Murashige and Skoog (MS) medium [34] (Duchefa, Netherlands, order number M0256) under 10 hours or 16 hours of light at 20–22°C and 65% humidity.

## Bioassay for Callose Deposition and Pathogen Inoculation

For callose induction, flg22 was applied at 1 μM for 24 hrs, and callose deposits were stained with aniline blue and visualized as described before [27]. Briefly, ten-day-old seedlings were transferred to liquid MS media with and without 1 μM flg22, destained after 24 hours in acetic acid - ethanol (1:3) for four hours, washed twice with $ddH_2O$, and incubated o/n in aniline blue solution (150 mM $KH_2PO_4$, 0.01% (w/v) aniline blue, pH 9.5). For infecting *A. thaliana* with the *Hpa*oomycete, suspensions of $5 \times 10^4$ spores/ml of *Hpa* strain Waco were spray-inoculated onto 14-day-old seedling and incubated at high humidity at 18°C as previously described [26]. Infected leaves were stained for callose 6 days post-inoculation.

## Microscopy and Image Acquisition

Stained callose was visualized using an ultraviolet epifluorescence microscope (Zeiss Axiophot, Carl Zeiss AG, Oberkochen, Germany). All microscope images were saved in TIFF or PNG format.

## Image Processing

After collecting callose images, we batch process images using CalloseMeasurer. Users are required to drag and drop the CalloseMeasurer script into the Acapella interface and then tick selection boxes according to their analysis requirements (e.g., image directory, image format, types of callose to measure, whether or not to construct spreading callose networks). After setting the input parameters, the analysis workflow can be initiated by clicking the "Run Script" button (Figure 5A). Following the batch processing, a set of PNG images with recognised callose deposits (coloured green) are generated. Features such as size, shape, and fluorescence signal intensity are measured based on every recognised callose deposit and saved in CSV files. If the "Detect Callose Network" option is selected, spreading callose networks are constructed (Figure 5B).

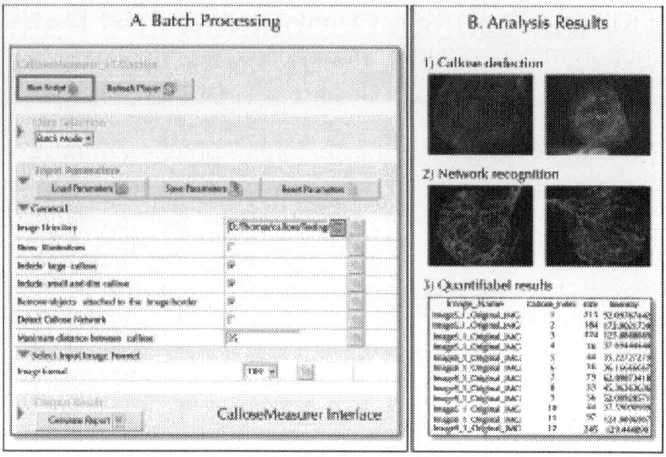


**Figure 5:** A generic workflow of batch processing flg22-induced callose deposits image sequence using CalloseMeasurer. (A) A series of callose images are batch processed by CalloseMeasurer (v1.0). (B) CalloseMeasurer detects callose deposits (highlighted in green) and identifies spreading callose networks (optional, coloured red). Generated images are saved as PNG files. Quantifiable callose features such as size, shape, fluorescence signal intensity, and area/length of spreading callose networks are exported to two CSV files for further statistical analysis.

# ACKNOWLEDGEMENTS

We like to thank Dr Richard Morris and Dr Matthew Hartley (John Innes Centre, Norwich Research Park) for critically reading the manuscript and members of the Robatzek laboratory for fruitful discussions and help.

Funding: T.S. was supported by a grant of the Deutsche Forschungsgemeinschaft (SPP1212), and research in the Robatzek laboratory is supported by the Gatsby Charitable Foundation.

# REFERENCES

1. Chen X-Y, Kim J-Y: Callose synthesis in higher plants. *Plant Signal Behav* 2009, 4:489-92.

2. Kudlicka K, Brown RM: Cellulose and Callose Biosynthesis in Higher Plants (I. Solubilization and Separation of (1->3)- and (1->4)-[beta]-Glucan Synthase Activities from Mung Bean). *Plant Physiol* 1997, 115:643-656.
3. Luna E, Pastor V, Robert J, Flors V, Mauch-Mani B, Ton J: Callose deposition: a multifaceted plant defense response.*Mol Plant Microbe Interact* 2011, 24:183-93.
4. Afzal AJ, da Cunha L, Mackey D: Separable fragments and membrane tethering of Arabidopsis RIN4 regulate its suppression of PAMP-triggered immunity. *Plant cell* 2011, 23:3798-811.
5. DebRoy S, Thilmony R, Kwack Y-B, Nomura K, He SY: A family of conserved bacterial effectors inhibits salicylic acid-mediated basal immunity and promotes disease necrosis in plants. *Proc Natl Acad Sci U S A* 2004, 101:9927-32.
6. Oh H-S, Park DH, Collmer A: Components of the Pseudomonas syringae type III secretion system can suppress and may elicit plant innate immunity. *Mol Plant Microbe Interact* 2010, 23:727-39.
7. Kim M, Mackey D: Measuring cell-wall-based defenses and their effect on bacterial growth in Arabidopsis. *Methods Mol Biol* 2008, 415:443-452.
8. Caillaud M-C, Piquerez SJM, Fabro G, Steinbrenner J, Ishaque N, Beynon J, Jones JDG:Subcellular localization of the Hpa RxLR effector repertoire identifies a tonoplast-associated protein HaRxL17 that confers enhanced plant susceptibility. *Plant J* 2012, 69:252-65.
9. Ham JH, Kim MG, Lee SY, Mackey D: Layered basal defenses underlie non-host resistance of Arabidopsis to Pseudomonas syringae pv. phaseolicola. *Plant J* 2007, 51:604-16.
10. Beck M, Zhou J, Faulkner C, Maclean D, Robatzek S: Spatio-temporal cellular dynamics of the Arabidopsis flagellin receptor reveal activation status-dependent endosomal sorting. *Plant Cell* 2012, 24:4205-4219.
11. Salomon S, Grunewald D, Stüber K, Schaaf S, MacLean D, Schulze-Lefert P, Robatzek S:High-throughput confocal imaging of intact live tissue enables quantification of membrane trafficking in Arabidopsis. *Plant physiol* 2010, 154:1096-104.

12. Fitzgibbon J, Beck M, Zhou J, Faulkner C, Robatzek S, Oparka K: A developmental framework for complex plasmodesmata formation revealed by large-scale imaging of the Arabidopsis leaf epidermis. *Plant cell Accepted* 2012, 1-22.

13. Collins T: ImageJ for microscopy. *BioTechniques* 2007, 43:S25-S30.

14. Schindelin J, Arganda-Carreras I, Frise E, Kaynig V, Longair M, Pietzsch T, Preibisch S, Rueden C, Saalfeld S, Schmid B, Tinevez J-Y, White DJ, Hartenstein V, Eliceiri K, Tomancak P, Cardona A: Fiji: an open-source platform for biological-image analysis. *Nature methods* 2012, 9:676-82.

15. Simpson C, Thomas C, Findlay K, Bayer E, Maule AJ: An Arabidopsis GPI-anchor plasmodesmal neck protein with callose binding activity and potential to regulate cell-to-cell trafficking. *The Plant cell* 2009, 21:581-94.

16. Cai G, Faleri C, Del Casino C, Emons AMC, Cresti M: Distribution of callose synthase, cellulose synthase, and sucrose synthase in tobacco pollen tube is controlled in dissimilar ways by actin filaments and microtubules. *Plant physiol* 2011, 155:1169-90.

17. de Chaumont F, Dallongeville S, Chenouard N, Hervé N, Pop S, Provoost T, Meas-Yedid V, Pankajakshan P, Lecomte T, Le Montagner Y, Lagache T, Dufour A, Olivo-Marin J-C: Icy: an open bioimage informatics platform for extended reproducible research. *Nat methods* 2012, 9:690-6.

18. Kimori Y, Baba N, Morone N: Extended morphological processing: a practical method for automatic spot detection of biological markers from microscopic images. *BMC bioinformatics* 2010, 11:373.

19. Pau G, Fuchs F, Sklyar O, Boutros M, Huber W: EBImage--an R package for image processing with applications to cellular phenotypes. *Bioinformatics* 2010, 26:979-81.

20. Sonka M, Hlavac V, Boyle R: *Image Processing, Analysis, and Machine Vision*. Second edition. Monterey, USA: PWS publishing; 1999:755.

21. Butenuth M: Segmentation of Imagery Using Network Snakes. *Int Arch Photogrammetry, Remote Sensing Spat Inf Sci* 2006, XXXVI:1-6.

22. Petrou M, Petrou C: *Image processing: the fundamentals*. second edition. Singapore: Wiley; 2010:794.
23. Gómez-Gómez L, Felix G, Boller T: A single locus determines sensitivity to bacterial flagellin in Arabidopsis thaliana. *Plant J* 1999, 18:277-284.
24. Ham JH, Majerczak D, Ewert S, SREEREKHA MV, Mackey D, Coplin D: WtsE, an AvrE-family type III effector protein of Pantoea stewartii subsp. stewartii, causes cell death in non-host plants. *Mol Plant Pathol* 2008, 9:633-643.
25. Gómez-Gómez L, Boller T: FLS2: an LRR receptor-like kinase involved in the perception of the bacterial elicitor flagellin in Arabidopsis. *Mol cell* 2000, 5:1003-11.
26. Tör M, Gordon P, Cuzick A, Eulgem T, Sinapidou E, Mert-Türk F, Can C, Dangl JL, Holub EB:Arabidopsis SGT1b is required for defense signaling conferred by several downy mildew resistance genes. *Plant Cell* 2002, 14:993-1003.
27. Lu Y-J, Schornack S, Spallek T, Geldner N, Chory J, Schellmann S, Schumacher K, Kamoun S, Robatzek S: Patterns of plant subcellular responses to successful oomycete infections reveal differences in host cell reprogramming and endocytic trafficking. *Cell microbiol* 2012, 14:682-97.
28. Niederlein A, Meyenhofer F, White D, Bickle M: Image analysis in high-content screening. *Combinatorial chem high throughput screening* 2009, 12:899-907.
29. Zanella F, Lorens J, Link W: High content screening: seeing is believing. *Trends Biotechnol* 2010, 28:237-45.
30. Eliceiri K, Berthold M, Goldberg I, Ibáñez L, Manjunath B, Martone ME, Murphy RF, Peng H, Plant AL, Roysam B, Stuurmann N, Swedlow JR, Tomancak P, Carpenter AE: Biological imaging software tools. *Nature methods* 2012, 9:697-710.
31. Peng H, Bateman A, Valencia A, Wren JD: Bioimage informatics: a new category in Bioinformatics. *Bioinformatics* 2012, 28:1057.
32. Zimmer C: From microbes to numbers: extracting meaningful quantities from images. *Cell Microbiol* 2012, 14:1828-1835.
33. Myers G: Why bioimage informatics matters *Nature methods* 2012, 9:659-60.

34. Murashige T, Skoog F: A revised medium for rapid growth and bioassays with tobacco tissue cultures. *Physiologia Plantarum* 1962, 15:473-497.

Chapter 6

# Delivering High-Resolution Landmarks Using Inkjet Micropatterning For Spatial Monitoring of Leaf Expansion

Lisheng Wang[1*], Simon T Beyer[1], Quentin CB Cronk[2] and Konrad Walus[1*]

[1]Department of Electrical and Computer Engineering, The University of British Columbia, 2332 Main Mall, V6T1Z4, Vancouver, BC, Canada

[2]Department of Botany and Faculty of Land and Food Systems, The University of British Columbia, 6270 University Blvd., V6T1Z4, Vancouver, BC, Canada

# ABSTRACT

## Background

Inkjet micropatterning is a versatile deposition technique with broad applications in numerous fields. However, its application in plant science is largely unexplored. Leaf expansion is one of the most important parameters in the field of plant science and many methods have been developed to examine differential expansion rates of different parts of the leaf lamina. Among them, methods based on the tracking of natural landmarks through digital imaging require a complicated setup in which the leaf must remain fixed and under tension. Furthermore, the resolution is limited to that of the natural landmarks, which are often difficult to find, particularly in young leaves. To study the fine scale expansion dynamics of the leaf lamina using artificial landmarks it is necessary to place small, noninvasive marks on a leaf surface and then recover the location of those marks after a period of time.

## Results

To monitor leaf expansion in two dimensions, at very fine scales, we used a custom designed inkjet micropatterning system to print a grid composed of c. 0.19 $mm^2$ cells on small developing leaves of ivy (*Hedera helix*) using 40 μm dots at a spacing of c. 91 μm. The leaves in different growing stages were imaged under magnification to extract the coordinates of the marks which were then used in subsequent computer-assisted leaf expansion analyses. As an example we obtained quantified global and local expansion information and created expansion maps over the entire leaf surface. The results reveal a striking pattern of fine-scale expansion differences over short periods of time. In these experiments, the base of the leaf is a «cold spot» for expansion, while the leaf sinuses are «hot spots» for expansion. We have also measured a strong shading effect on leaf expansion. We discuss the features required to build an inkjet printing apparatus optimized for use in plant science, which will further maximize the range of tissues that can be printed at these scales.

## Conclusions

To apply inkjet micropatterning to plant studies, we have successfully delivered landmarks on ivy leaf surfaces and achieved high-resolution, two-dimensional monitoring of leaf expansion at different growing stages. The measurement is capable of reliably identifying the fine scale changes during plant growth. As well as delivering landmarks, this technology may be used to deliver microscale targeted biological components such as growth hormones, and possibly be used to pattern sensors directly on the leaves.

# BACKGROUND

Inkjet printing is a technique involving ejecting tiny liquid droplets in a non-contact manner onto target object surfaces to form any desired patterns in high resolution. Historically, it has been used almost exclusively for document and image printing, finding considerable success in the consumer electronics market. More recently the potential of this technology has been explored to a greater degree, especially in developing printable electronics. Popular applications include polymer solar cells, thin film transistors, sensors and organic light emitting diode arrays [1]. If functional materials and patterns are printed directly on plant bodies, various plant behaviours can be accurately investigated in exquisite detail which raises the possibility of "sentinel plants" reporting on growth conditions (desiccation, gas exchange, photon flux density etc.). The non-contact nature of inkjet printing makes it a suitable tool for depositing materials on plant surfaces without causing physical damage. Therefore, we consider the application of inkjet printing to plant studies is not only of great significance but also very feasible. However, this area has scarcely been explored until now.

Among various topics in botany, precise understanding of the growth of plant organs is important not only in the study of morphogenesis but also in the applied plant sciences, as the expansion of the leaf lamina is an important factor contributing to plant yield. Change in total leaf area is easy to measure as the leaf outline is easily captured. However, differential expansion of different parts of the leaf is more difficult to study as natural landmarks within the leaf are difficult to find.

The pattern of veins can provide landmarks, particularly if the venation is finely reticulate, but at early stages the veins are poorly formed and the pattern changes rapidly. Nevertheless, for later stages of leaf development the question of whether various parts of the lamina expand at the same or at different rates throughout development have been answered using the expansion of small areas delineated by veins [2]. However, this method cannot collect measurements with a high spatial resolution.

In a more recent study by Walter *et al.*[3], leaf venation was tracked using advanced digital image sequence processing, resulting in a method with very high temporal resolution. However, the spatial resolution of this method and the requirement that the leaves remain mechanically fixed to the apparatus during growth makes it more suitable for large leaves.

Finer scale measurements can be achieved with clonal analysis in which individual cell lineages are followed using phenotypic markers [4-7] but this depends on the availability of suitable biological material. It has successfully been used to study leaf and petal expansion using visible changes mediated by, for instance, transposon activity.

Another method involves placing artificial landmarks on the plant organ. Avery [8], in a pioneering study of leaf expansion, marked a grid on a developing tobacco leaf using India ink. This technique is only suitable for relatively large leaves, typically tobacco leaves c. 6 cm long, and using a fairly coarse initial landmark spacing (c. 5 mm). Numerous subsequent authors have used the same technique [9-11]. A modern variant of this procedure is to use graphite particles scattered randomly[12]. This technique can reach much finer scales but requires careful image analysis to reconstruct the landmarks as they move apart during expansion.

To address some of these problems of artificial landmarks, Hamamoto, Kimura & Yamaguchi [13] used non-contact inkjet printing to print lines of small dots (0.1-0.2 mm in diameter) on the surfaces of plant organs. However, the authors only studied the elongation of the organs along one direction. The purpose of this paper is to extend the inkjet micropatterning method, printing at much finer scales and in two dimensions, using 40 μm dots, in order to study the expansion of very young leaves. We used ivy (*Hedera helix* cv) as the experimental material. The complex outline of its palmate-shaped leaves allowed us to study the role of differential expansion in the morphogenesis of such complex shapes.

# METHODS

## Plant Material and Growth

The plant used in the experiments was English ivy (*Hedera helix* cv) shown in Figure 1. It was grown in a 20 cm-diameter pot positioned with side illumination from a window and kept at a relatively constant temperature of c. 25 °C. The plant was able to receive direct sunlight from the southeast for approximately five hours per day throughout the experimental period (summer). It was watered two to three times per week and supplied with nutrients using houseplant nutrient spikes (Miracle-Gro®, The Scotts Co.) once before the experiment to ensure that leaf growth was unaffected by nutrient or water deficits.

**Figure 1:** Photograph of studied ivy plant. The branches with studied leaves were supported at the top of the plant and separated from others.

The ten studied leaves were selected from ten primary branches of the plant. Each leaf was 17-20 mm long at Day 0 of the experiment, corresponding to a leaf plastochron index (LPI) of c. 0.8-0.9 [2] assuming that the LPI of a 10 mm long leaf is set at zero by convention. The leaves were measured at the 5th, 10th and 15th day, corresponding to a growth interval of c. 0.3, 0.6 and 0.9 plastochrons respectively (where one plastochron is the interval between the production of successive leaves on a shoot). The branches bearing the leaves under study were supported at the top of the plant and separated from others to make sure they received ample natural light distributed as evenly as possible (Figure 1).

## Pattern Printing on Leaf Surfaces

The inkjet printing was performed using a custom-built drop-on-demand micropatterning system. Figure 2 shows the schematic diagram of this system. We used a 40 μm orifice diameter piezoelectrically-driven nozzle (MJ-AB-63-40; MicroFab Technologies, Inc.). When actuated with a voltage pulse, the nozzle was able to eject droplets of ink on the order of picoliters per drop, forming individual dots on the leaves. The voltage level used for actuation was 29 V with a 92 μs wide sinusoidal pulse. A syringe pump (PSD/3 5175-01; Hamilton Inc.) combined with an inline pressure sensor (PX139-001D4V; OMEGA, Inc.) was used to control the air pressure in the ink canister in order to maintain consistent droplet formation as ink was consumed.

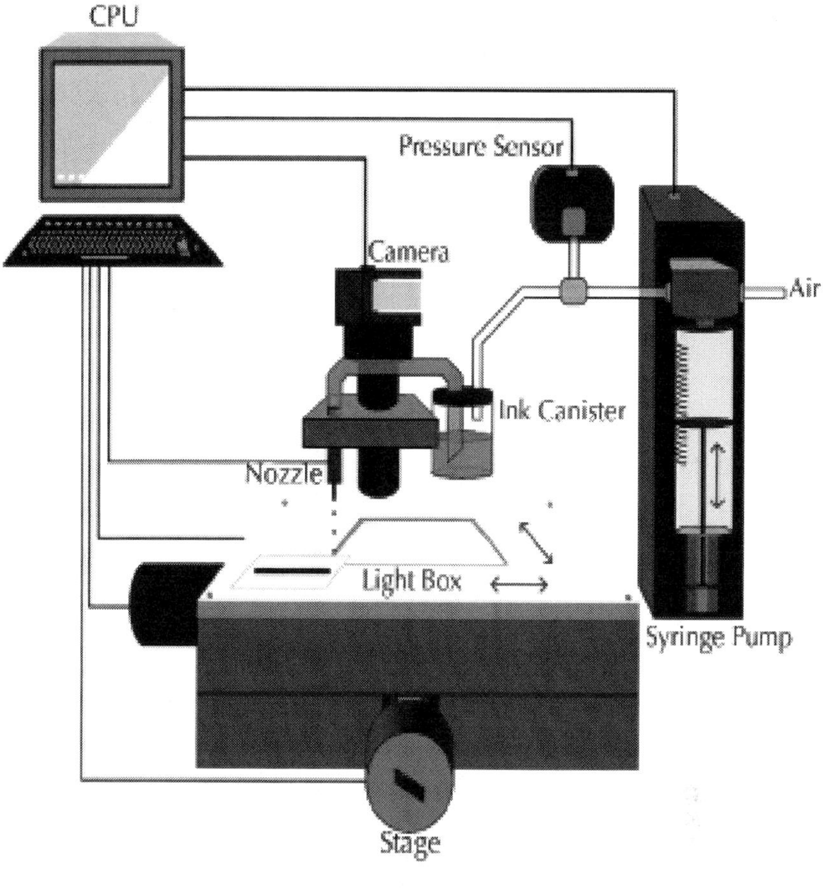


**Figure 2:** Schematic diagram of the inkjet printing/imaging system. The inkjet printing system includes a nozzle, pressure sensor, ink canister, syringe pump, CPU and stage, and the imaging system includes a camera, light box, CPU and stage.

To reduce nozzle vibration, the nozzle was kept stationary while the deposited pattern was formed through the movement of a two-dimensional stage. Stage movement with resolution of 2.5 μm was accomplished through the use of two stepper motors (23D - 6306C; American Precision Industries). Each leaf, along with the branch, was fastened onto the stage before printing by folding a piece of weakly adhesive tape, applying it to the back of the leaf and gently pressing to flatten it. The ink used was violet food colouring (Scott Bathgate Ltd.) which provided distinguishable contrast to the green leaves.

All components of the inkjet printer were controlled with a PC hosting custom C++ based software. Macro based commands were used to determine the patterns to be printed on the leaves. The pattern used was a square grid consisting of 56 horizontal lines and 56 vertical lines; sufficient to cover the entire surface of each leaf. Each line was composed of individual dots with a centre-to-centre distance of 87 μm. Line spacing in both the vertical and horizontal directions was 435 μm.

## Pattern Recording

The recording of printed patterns was carried out on the same 2D stage (Figure 2). A USB CMOS camera (EO-1312M; Edmund Optics, Inc.) with a 0.42× zoom lens (HRP042-CMT; Diagnostic Instruments, Inc.) was mounted beside the nozzle and connected to a computer. Uc480Viewer (Thorlabs, Inc.) was used to capture digital images. A home-built LED light box was attached to the top of the stage to provide backlighting during imaging. Each leaf was placed on the window of the light box and gently covered with a glass slide to fix and flatten its surface. The stage was stepped in a grid pattern to produce a total of 9-16 photographs (depending on the size of the leaf) of different regions of the leaf. A panoramic image was then created by stitching those photographs together using Windows Live Photo Gallery 2009 (Microsoft Co.).

## Pattern Extraction and Computer-Assisted Analysis

The method and algorithm to acquire quantitative and visualized expansion information of one leaf is composed of the following steps:

1. Manually rotate the leaf image with the initial printed pattern until the printed lines are approximately parallel to the horizontal and vertical axes of the screen.
2. Open the rotated image file with ImageJ 1.43j (National Institute of Health of United States; NIH). Manually select the intersecting points of the crossing lines, as well as the intersecting points of the printed lines and the leaf margin, using ImageJ to store the coordinates. If two or more points are very close to each other,

select only one of them. Save the index numbers and coordinates of all selected points in a text file and those of margin points again in a separate text file.

3. The data analysis program, written using Matlab R2008a (The Mathworks, Inc.), is used to import the text files. The program takes each point from the text file and locates its neighbouring points by searching for the nearest one along the ascending *x* and *y* axis respectively as well as any point in its first quadrant to which the distance is no more than 650 μm (a little longer than times line spacing). Thus, the program will find three neighbours for points within the leaf and 0-4 neighbours for points along the leaf margin. The program discards those points with no more than one neighbour as well as those without neighbours along both the ascending ¬*x* and *y* axes. It is worth mentioning that the high accuracy of the inkjet printing process makes automatic selection of neighbouring points a feasible possibility.
4. Each point and its neighbouring points specify a small square or triangular cell. Here, and for the remainder of this paper, a cell is defined as the region between printed lines rather than in the biological sense. The program records the index numbers of those points and calculates the area of the cell ($A_0$) using the *polyarea* function. It also discards those cells which are outside of the leaf surface (specified by the recorded margin points) using the *inpolygon* function. The surface area of the entire leaf can be calculated either based on the polygon specified by the margin points or as the summation of the areas of all cells. Further study shows the difference between these two values of all studied leaves is less than 3%, which is negligible.
5. For distorted patterns at Day *x* after leaf expansion, repeat Steps 1 and 2. Select the points in the exact same order as in Step 2. That is, each point is expected to have the same serial number in the data files before and after leaf expansion.
6. The program calculates the areas of expanded cells (*Ax*) specified in Step 4 as well as total cell expansion (*Ex*) and relative expansion rates (*GRx-y*), defined as:

$$E_x = \frac{A_x}{A_0} \tag{1}$$

$$GR_{x-y} = \frac{A_y - A_x}{A_x(y-x)} \tag{2}$$

7. Lastly the program creates a simple leaf expansion map using the *patch* function and a contour map using the *contourf* function.

Computer programs and scripts used in this study are available upon request.

# RESULTS

## Pattern Printing and Data Acquisition

A total of ten young actively growing ivy leaves were selected, which were of similar developmental age and size at the start of the experiments (Table 1). We used a custom inkjet printer for non-contact printing on six of the leaves (Leaves 1-6) using commercial food colouring as the printing material.

**Table 1:** Size information, expansion and relative expansion rate of leaves at different time periods

| Leaf No. | Initial size | | Expansion ($mm^2/mm^2$) | | | Relative expansion rate ($mm^2/mm^2/day$) | | | Experimental details |
|---|---|---|---|---|---|---|---|---|---|
| | Length (mm) | Area ($mm^2$) | Day 5 | Day 10 | Day 15 | Day 0-5 | Day 5-10 | Day 10-15 | |
| 1 | 20 | 184 | 2.7 | 3.5 | 3.7 | 0.33 | 0.064 | 0.012 | Patterned. |
| 2 | 18 | 159 | 2.6 | 3.4 | 3.7 | 0.32 | 0.065 | 0.013 | |
| 3 | 20 | 214 | 1.7 | 1.8 | 1.9 | 0.13 | 0.019 | 0.007 | |
| 4 | 20 | 195 | 1.5 | 1.6 | 1.6 | 0.10 | 0.021 | 0.006 | |
| 5 | 17 | 169 | 1.3 | 1.4 | 1.4 | 0.05 | 0.019 | 0.008 | Patterned; partly covered. |
| 6 | 18 | 202 | 1.4 | 1.6 | 1.7 | 0.08 | 0.034 | 0.010 | |
| 7 | 19 | 176 | 1.9 | 2.2 | 2.3 | 0.17 | 0.038 | 0.009 | Controls. |
| 8 | 19 | 180 | 1.5 | 1.7 | 1.7 | 0.10 | 0.020 | 0.006 | |
| 9 | 20 | 193 | 2.3 | 2.7 | 2.8 | 0.25 | 0.040 | 0.007 | |
| 10 | 17 | 163 | 2.5 | 3.3 | 3.5 | 0.30 | 0.067 | 0.011 | |

As shown in Figure 3, each leaf is marked with a high-resolution square grid. Each grid line is patterned with discontinuous dots of c. 40 μm in diameter. The average distance between two adjacent dots is 91 μm and the average spacing between two parallel lines is 435 μm (Inset of Figure3). The entire surface of each leaf is divided into a grid of 700-1000 small cells depending on the size of the leaf. Most cells are located on the planar portions of the leaves and are approximately square-shaped with an average area of 0.19 $mm^2$ (region 'a' in inset of Figure 3). Some cells are somewhat deformed because of the curved surface (Region 'b' in inset of Figure 3) and those near the margin may have other shapes, e.g. triangles, trapezoids or pentagons, etc. (Region 'c' in inset of Figure 3). Fortunately, the shape of the cells does not influence the quantitative expansion analysis since we are measuring their relative expansion.

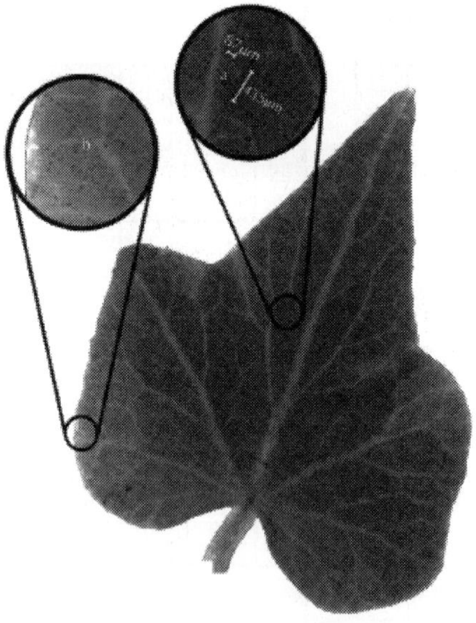

**Figure 3:** One ivy leaf (Leaf 4) with printed grids on its surface. Each grid line is composed of discontinuous dots. The entire surface is divided into 700 small cells. Regions a (square), b (deformed square) and c (triangle) represent different shapes of the cells.

To study the influence of the inkjet printing process on leaf viability, four additional leaves were selected as unprinted controls (Leaves 7-10).

The studied leaves were monitored three times at five day intervals throughout the experiment. In other experiments, more frequent measurements over longer intervals could certainly be available if necessary. The patterned leaves were non-destructively imaged and the patterns were extracted and digitized manually for analysis. The intersection points between crossing lines and between printed lines and leaf margins were marked and their coordinates recorded. Figure 4 shows 25 example points (white circles) and two marker-defined cells (Cell A and Cell B). Those recorded coordinate values from each measurement are fundamentals in our monitoring system, with which we are able to perform various analyses using different models of interest. The following leaf expansion quantification, global expansion mapping and contour plotting are three demonstrating examples.

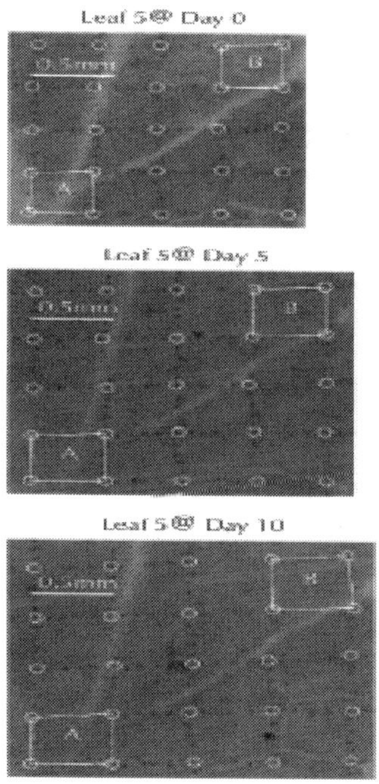


**Figure 4:** A small region of Leaf 5 at different growing stages showing the change of printed grid. 25 example points (white circles) and two marker-defined cells (Cell A and Cell B) were labelled. Each cell does not experience the same rate of expansion. Cell A is 1.37 times larger at Day 10 than at Day 0 while Cell B is 1.62 times larger.

## Quantifying Leaf Expansion

With the acquired coordinates, we are able to calculate the area of each marker-defined cell. By recording the displacement of each printed point and the expansion of each cell, we were able to track the local expansion patterns of each leaf.

As an example, Figure 4 shows the morphology of a small 16 cell region of Leaf 5 over a 10 day span. The initial area of this region was 3.08 $mm^2$. At Day 5 and Day 10 the total area increased to 4.05 $mm^2$ (1.3 times) and 4.57 $mm^2$ (1.5 times) respectively. From the image it is clear that each cell does not experience same rate of expansion. For example, Cell A in Figure 4 is 1.37 times larger at Day 10 than at Day 0 while Cell B is 1.62 times larger. This difference demonstrates the highly local expansion differences on the leaf.

The non-uniformity of leaf expansion is also evident in the deformation of the printed grid. In Figure4, the top two horizontal lines are parallel to the bottom horizontal line at Day 0, but by Day 10 they are clearly no longer parallel. We believe this is due to a faster expansion rate in the upper right region of the leaf.

We are also able to determine the overall expansion of one leaf by tracking the surface area increase of the entire leaf. The average expansion of the patterned leaves (Leaves 1-4) is 2.1 at Day 5, 2.5 at Day 10 and 2.6 at Day 15 while those of controls (Leaves 7-10) are 2.0 at Day 5, 2.5 at Day 10 and 2.6 at Day 15. The patterned leaves grew at a similar rate as the controls, suggesting that the inkjet printing process does not have any significant influence on the expansion of the leaves.

The mean relative expansion rates at different time periods during the experiment (Table 1) show the transition from fast expansion to slow expansion as the leaf nears maturity. As seen in Figure 5 the relationship between the expansion rate and the observation day is approximately log-linear with an average correlation coefficient of 0.996 among the ten leaves studied. This suggests that the expansion rate of the leaves decreases exponentially during this phase. More frequent measurements could be used to gain a precise quantification of expansion rate changes over time.

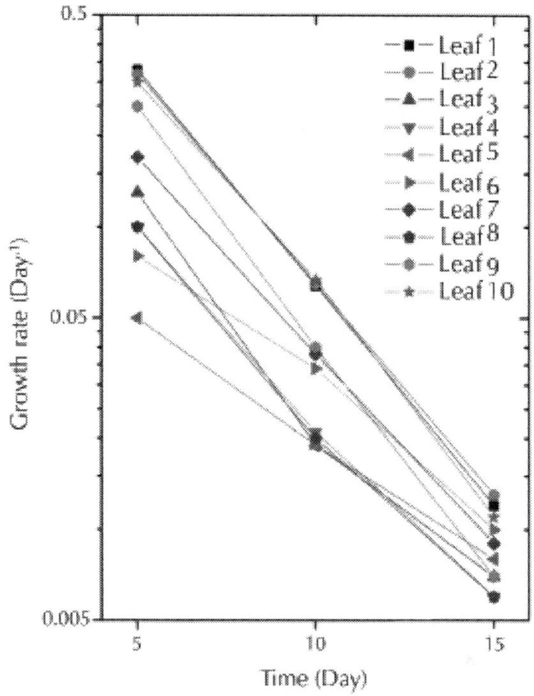


**Figure 5:** Relationship between the observation date and expansion rate of ten studied leaves. The relationship is approximately log-linear with an average correlation coefficient of 0.996 among the ten leaves studied.

## Mapping of Leaf Expansion

By colouring the marker-defined cells according to their expansion values we are able to produce maps showing the two-dimensional information of leaf expansion. The colour map, from blue, to red, indicates an increase in the rate of expansion.

Figure 6 shows the expansion maps for Leaves 1-3 at Day 5 and Day 10. The local expansion of the leaves is clearly expressed through the colour variation of each cell. Generally speaking, Leaves 1 and 2 have similar expansion rates throughout the entire experiment, both of which are larger than Leaf 3. In addition, the expansion of Leaves 1 and 2 is generally bilaterally symmetrical while Leaf 3 shows a certain degree of asymmetry in its expansion.

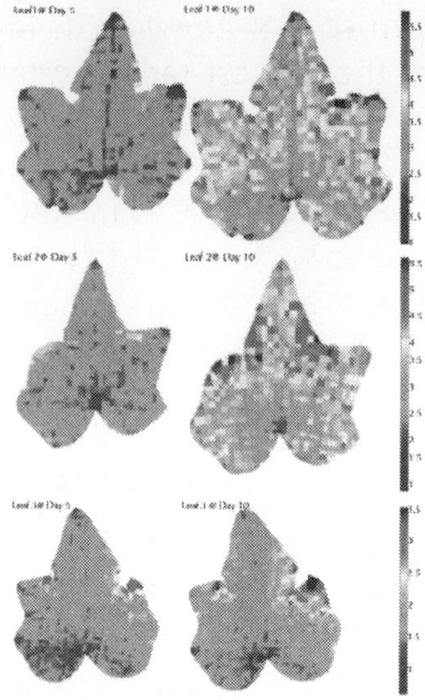

**Figure 6:** Mapping of Leaves 1-3 in different growing stages. The local expansion of the leaves is expressed through the color variation of each cell. The base of each leaf has the smallest degree of expansion and the sinus areas grow the fastest. Unlike Leaves 1 and 2, Leaf 3 shows a certain degree of asymmetry in its expansion.

The non-uniform expansion of the leaves is clearly demonstrated by the expansion maps. We can see that the base of each leaf has the smallest degree of expansion, indicating this area consistently grows the slowest, called "cold spot". In contrast, the fastest expansion consistently occurs around the sinus areas between the terminal lobe and lateral lobes, called "hot spots", with an average expansion up to 3 times larger than the base area at Day 10. It is also clear from the results that the expansion rates along the primary veins, especially the midribs, are slower compared to their surrounding areas.

The expansion rate of each tip area is significantly correlated with the shape of that tip. A slower expansion rate will lead to a sharp tip, like most tips of the three leaves. For example, the tips of the terminal lobes show smaller expansion than their surrounding areas in all trials. In contrast, a faster expansion rate leads to a round-shaped tip.

To the best of our knowledge, spatial and temporal patterns of ivy leaf growth have not been reported before. Existing literature suggests that dicot leaves either grow with a base to tip growth gradient, or with a relatively uniform spatial growth [3]. In our results the sinus regions appear to have the highest rate of growth and the leaf base and major veins appear to have the slowest, which could be seen as a departure from the growth patterns of other studied dicot leaves. However, the gradual bending of the sinus regions of our ivy leaves over the experimental period is indicative of a higher rate of growth there, and partly validates the findings described above.

To further verify the validity and accuracy of our method, we intentionally covered part of two leaf surfaces (Leaves 5-6) with aluminium foil which would supposedly lead to asymmetric leaf expansion due to uneven illumination. Figure 7 shows the mapping results of both leaves at Days 5 and 10. It is very clear that the uncovered regions are expanding more quickly (mainly yellow and red) than the covered regions (blue) at both Day 5 and 10 leading to the absence of bilateral symmetry in the expansion patterns which commonly exists in normal leaves (Leaves 1-2).. Quantitatively, the overall expansion of the uncovered region is 1.4 and 1.6 times larger than the covered region at Day 10 in Leaves 5 and 6, respectively. Such a result is consistent with our hypothesis and in accordance with Sue et al., who characterized the expansion patterns of normal and malformed *Vitis vinifera* cv. *Ruby Red* leaves [14].

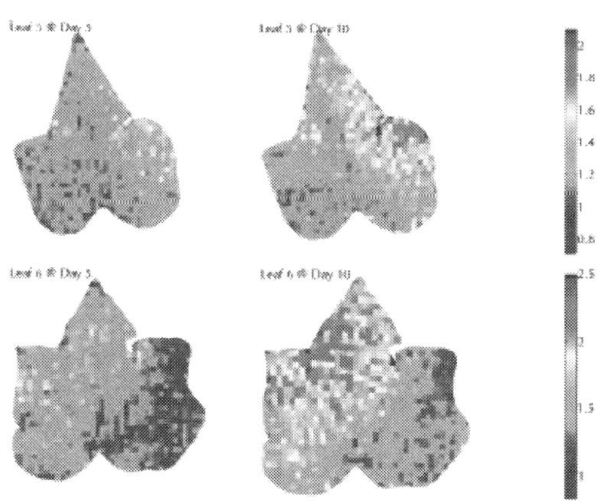

**Figure 7:** Mapping of Leaves 5-6 in different growing stages. Both leaves were partly covered using alumina foil, including the left side from the midrib of Leaf 5 and the right basal and lateral lobes of Leaf 6. The uncovered regions are expanding more quickly than the covered regions at both Day 5 and 10.

Contour plotting is another method to describe the expansion difference over an entire leaf surface. Figure 8(B) shows the colour-filled contour map of Leaf 2 at Day 10. Similar to the corresponding simple map (Figure 8A), it also shows the base to be the slowest expansion region and sinuses to be the fastest. The relationship between tip shapes and tip expansion rates are also demonstrated. The slower expansion along the veins (marked by black dots inside the map) becomes more visible in the contour map.

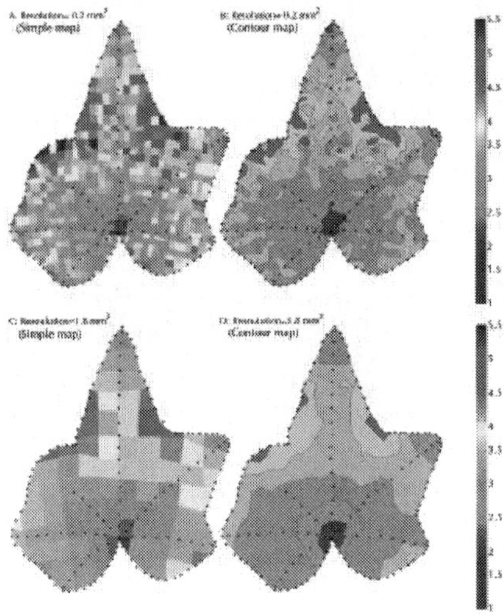

**Figure 8:** Comparison of simple mapping and contour plotting with different printing resolutions. Comparisons are made on Leaf 2 at Day 10. Black dots outlined the margin and major veins of the leaf. Contour plotting provides similar information to simple mapping with the expansion variation between regions more continuous. In lower-resolution maps, some important details are missing.

## Error Analysis

As explained above, the 2D mapping of leaf expansion is comprised of three steps: pattern printing, optical recording and digital extraction. To study the accuracy and resolution of this technique, we performed error analysis on a patterned region containing 224 cells (including 190 square cells) for each of the three steps.

First, we determined the spatial resolution of the CCD camera and lens setup by finding the number of pixels per millimetre in an image with known dimensions. We found the resolution to be 86 pixels/mm, or 11.6 μm/pixel. This is thus the lower resolution limit for this method of expansion measurement given the camera used.

Because the digital pattern extraction process involves manually clicking on the image with a mouse, there is some degree of human error. To investigate this error we performed the extraction process twice on one image, resulting in two sets of 265 dot coordinates. First, the positional deviation between corresponding dots along the X and Y directions was calculated. Statistical results (Figure9A) show that the error values (14.0 μm and 12.6 μm along the X and Y directions respectively) are larger than the spatial resolution of the images. Next we compared the areas of 224 cells from the two data sets and found a mean deviation of 1.00 and a standard error of 0.04 (Figure 9B). These results indicate that while the positional error is noticeable, the resulting error in cell area is negligibly small.

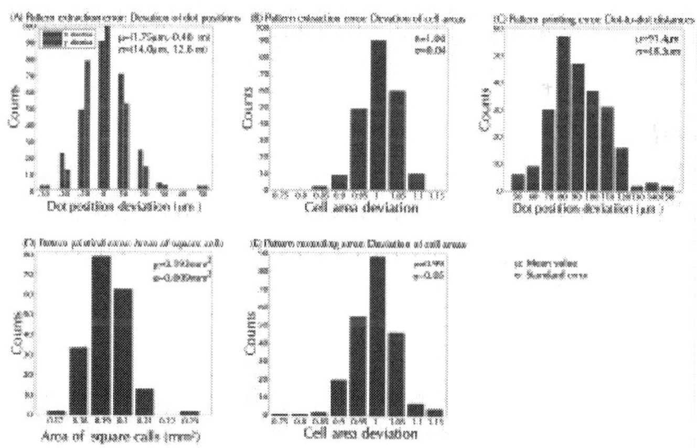


**Figure 9:** Error analysis of pattern extraction, printing and recording processes.

The accuracy of the physical printing process was determined in two steps. First, 240 pairs of adjacent printed dots were randomly chosen and their distances measured. The average distance was 91.4 μm with a standard error of 18.3 μm including the pattern extraction error. Figure 9C shows the distribution of the dot-to-dot distances. 91.4

μm is slightly larger than the expected dot distance of 87 μm, and 18.3 μm is substantially larger than the spatial resolution of the image. We suspect this error is due to inaccuracy in the movement of the stage, or the migration of the aqueous ink on the hydrophobic surface of the leaf. Although for this project the absolute positions of single dots are not important, precise study of surface properties and careful selection of inks would be necessary if more accurate printing is needed. The areas of a total of 190 square cells throughout the investigated region were also measured and their distribution analysed (Figure 9D). The average value is 0.19 $mm^2$ with a standard error of 0.01 $mm^2$, which can be neglected.

Lastly there is some degree of error associated with imaging the leaves due to the physical action of setting up the camera and loading the sample. To measure this error we imaged a printed pattern twice, reloading the sample and reassembling the camera setup each time. We found the deviation in the areas of 224 cells to have a mean value of 0.99 and a standard error of 0.05 (Figure 9E). This error is the combination of both the pattern generation and recording processes. Through this analysis we have found that the total error is substantially smaller than both the overall leaf expansion (Table 1) and the expansion of individual cells (Figures 6 and 7), indicating that error does not have a significant impact on the results of this method.

# DISCUSSION

Inkjet printing can deliver landmarks at fine scales and fast speeds. In the case of this research, by using 40 μm dots, a grid containing one thousand 0.2 $mm^2$ cells could be precisely printed in 10 min. Finer-scale printing can also be achieved using this method. This is a considerable improvement over the 25 $mm^2$ grids manually printed or stamped on leaves in classical growth studies and can potentially lead to the development of more sophisticated growth models encompassing a greater range of organs. In addition, because there is no contact between the nozzle and the organ surfaces in our custom built printer the plant organs are not mechanically damaged, as demonstrated by the control experiment described before.

To further study the influence of printing resolution on the measurements, we increased the grid cells from 0.2 $mm^2$ to 1.8 $mm^2$

in size by evenly selecting one out of three marker dots along both horizontal and vertical printed lines. The standard errors of deviation of cell areas are 0.01 during the pattern extraction process and 0.02 during the pattern recording process, both of which are substantially smaller than those of previously used finer grid (Figures 9B, 9E), indicating that a higher resolution would result in an increased measurement error. The mapping and contour plotting results are shown in Figure 8(C, D). Compared to the higher-resolution mapping results (Figure 8(A, B)), the "hot spots" and "cold spots" are also visible for expansion within one leaf, and the expansion variation from "hot spots" to "cold spots" becomes more continuous, virtually no isolated spots(Figure8A, B). However, due to the reduced resolution some important details are missing. As an example, the slower expansion rate along the veins is not visible. In addition, the expansion disparity between "hot spots" and "cold spots" is underestimated. Since high-resolution printing provides additional detail while potentially increase measurement error, the choice of an appropriate resolution must be made according to the requirement of each study.

The main limitations of the system described here are leaf access and surface suitability. Young leaves are often protected by older leaves and leaf structures such as stipules. Maneuvering the shoot tip so that young leaves are accessible to the inkjet printer is a significant challenge. Our current printing system has a fixed head and moving stage and the machine as a whole cannot be tilted to access harder to reach parts of a plant. Climbing plants like ivy are therefore ideal as their shoots can be introduced at any orientation into the printing chamber. Future innovations should include redesign of the printer to provide adjustable stage orientation to give greater flexibility with regard to printable plant material. Many young leaves are covered with dense hairs and are rolled or folded into one of several patterns of space packing or "ptyxis" making these measurements challenging. With our present system evenly spaced landmarking can only be achieved if the leaf is relatively flat. We flatten leaves with gentle pressure, holding them in place using a weakly adhesive surface. This is often possible with leaves that are folded along the midrib (conduplicate). However, leaves that are tightly rolled (supervolute) when young are not treatable in this way and only the exposed surfaces can be landmarked. In addition, leaf flattening is also required during the pattern recording process, as in many other existing techniques. However, leaves of many species will experience significant distortion during expansion, leading to undulating surfaces which are difficult to be flattened and imaged.

In this study we used an opaque mark consisting of food colouring that provided sufficient contrast when transilluminated. However, the use of fluorescent markers would potentially be a way of gaining greater contrast for use with automated image analysis which would speed the process of tracking the landmarks.

We use non-contact inkjet printing here to place landmarks on leaves. However, the versatility of inkjet technology enables it to be used as a delivery system for a wide range of interesting components [15,16]. Examples relevant to plant science might include phytohormones delivered in precisely quantifiable amounts (perhaps along with markers) as quantitative bioassays. Besides dye, inkjet technology can be used as a particle delivery system. Increasingly, sensor technology is being reduced in scale [17,18] and small nanoparticles consisting of indicator chemicals can in theory be used to monitor physicochemical conditions on the leaf surface.

# CONCLUSIONS

In this article, we demonstrate the applicability of inkjet micropatterning to the study of plant growth and achieved high-resolution and two-dimensional monitoring of leaf expansion. Mapping results show that this method is able to describe the expansion difference at different regions of a leaf lamina on a very fine scale. The measurement is reliable according to error analysis and control experiments suggest that this technique can reflect the change of conditions during plant growth. As well as delivering landmarks, this technology could be used to deliver microscale targeted biological chemicals such as growth hormones, and possibly be used to pattern sensors directly on the leaves.

# ACKNOWLEDGEMENTS

We acknowledge the financial assistance of a Martha Piper Research Award (UBC). The laboratories of Konrad Walus and Quentin Cronk are funded by the Natural Sciences and Engineering Research Council of Canada (NSERC) Discovery Grants Program. Gabriel Man, Christine Woollacott and Julia Nowak provided helpful assistance and discussion at an early stage in the development of this work.

# REFERENCES

1. Tekin E, Smith PJ, Schubert US: Inkjet printing as a deposition and patterning tool for polymers and inorganic particles. *Soft Matter* 2008, 4:703-713.
2. Maksymowych R: *Analysis of leaf development.* Cambridge [Eng.]: University Press; 1973.
3. Walter A, Silk WK, Schurr U: Environmental Effects on Spatial and Temporal Patterns of Leaf and Root Growth. *Annual Review of Plant Biology* 2009, 60:279-304.
4. Dolan L, Poethig RS: Clonal analysis of leaf development in cotton. *American Journal of Botany* 1998, 85:315-321.
5. Poethig RS, Sussex IM: The cellular-parameters of leaf development in tobacco - a clonal analysis. *Planta* 1985, 165:170-184.
6. Poethig RS, Szymkowiak EJ: Clonal analysis of leaf development in maize. *Maydica* 1995, 40:67-76.
7. Rolland-Lagan AG, Bangham JA, Coen E: Growth dynamics underlying petal shape and asymmetry. *Nature* 2003, 422:161-163.
8. Avery GS: Structure and development of the tobacco leaf. *American Journal of Botany* 1933, 20:565-592.
9. Poethig RS, Sussex IM: The developmental morphology and growth dynamics of the tobacco leaf. *Planta* 1985, 165:158-169.
10. Erickson RO: Relative elemental rates and anisotropy of growth in area - a computer programme. *Journal of Experimental Botany* 1966, 17:390-403.
11. Granier C, Tardieu F: Spatial and temporal analyses of expansion and cell cycle in sunflower leaves - A common pattern of development for all zones of a leaf and different leaves of a plant. *Plant Physiology* 1998, 116:991-1001.

12. Basu P, Pal A, Lynch JP, Brown KM: A novel image-analysis technique for kinematic study of growth and curvature. *Plant Physiology* 2007, 145:305-316.
13. Hamamoto H, Kimura M, Yamaguchi I: High resolution analysis of plant organ growth using non-contact inkjet marking. *Environmental and Experimental Botany* 2006, 57:62-69.
14. Wolf SD, Silk WK, Plant RE: Quantitative Patterns of Leaf Expansion - Comparison of Normal and Malformed Leaf Growth in Vitis-Vinifera Cv Ruby Red. *American Journal of Botany* 1986, 73:832-846.
15. de Gans BJ, Duineveld PC, Schubert US: Inkjet printing of polymers: State of the art and future developments. *Advanced Materials* 2004, 16:203-213.
16. Mabrook MF, Pearson C, Petty MC: An inkjet-printed chemical fuse. *Applied Physics Letters* 2005, 86:013507.
17. Mabrook MF, Pearson C, Petty MC: Inkjet-printed polypyrrole thin films for vapour sensing. *Sensors and Actuators B-Chemical* 2006, 115:547-551.
18. Li B, Santhanam S, Schultz L, Jeffries-El M, Iovu MC, Sauve G, Cooper J, Zhang R, Revelli JC, Kusne AG, *et al.*: Inkjet printed chemical sensor array based on polythiophene conductive polymers. *Sensors and Actuators B-Chemical* 2007, 123:651-660.

Chapter 7

# Rapid Metabolic Profiling of Nicotiana tabacum Defence Responses against Phytophthora Nicotianae Using Direct Infrared Laser Desorption Ionization Mass Spectrometry and Principal Component Analysis

Alfredo J Ibáñez[1], Judith Scharte[2], Philipp Bones[2], Alexander Pirkl[1], Stefan Meldau[3], Ian T Baldwin[3], Franz Hillenkamp[1], Engelbert Weis[2] and Klaus Dreisewerd[1*]

[1]Institute of Medical Physics and Biophysics, Westfälische Wilhelms-Universität Münster, Robert-Koch-Str. 31, D-48149 Münster, Germany

[2]Institute of Botany, Westfälische Wilhelms-Universität Münster, Schlossgarten 3, D-48149 Münster, Germany

[3]Department of Molecular Ecology, Max Planck Institute for Chemical Ecology, Beutenberg Campus, Hans-Knöll-Str. 8, D-07745 Jena, Germany

# ABSTRACT


## Background

Successful defence of tobacco plants against attack from the oomycete *Phytophthora nicotianae* includes a type of local programmed cell death called the hypersensitive response. Complex and not completely understood signaling processes are required to mediate the development of this defence in the infected tissue. Here, we demonstrate that different families of metabolites can be monitored in small pieces of infected, mechanically-stressed, and healthy tobacco leaves using direct infrared laser desorption ionization orthogonal time-of-flight mass spectrometry. The defence response was monitored for 1 - 9 hours post infection.

## Results

Infrared laser desorption ionization orthogonal time-of-flight mass spectrometry allows rapid and simultaneous detection in both negative and positive ion mode of a wide range of naturally occurring primary and secondary metabolites. An unsupervised principal component analysis was employed to identify correlations between changes in metabolite expression (obtained at different times and sample treatment conditions) and the overall defence response.

A one-dimensional projection of the principal components 1 and 2 obtained from positive ion mode spectra was used to generate a Biological Response Index (BRI). The BRI obtained for each sample treatment was compared with the number of dead cells found in the respective tissue. The high correlation between these two values suggested that the BRI provides a rapid assessment of the plant response against the pathogen infection. Evaluation of the loading plots of the principal components (1 and 2) reveals a correlation among three metabolic cascades and the defence response generated in infected leaves. Analysis of selected phytohormones by liquid chromatography electrospray ionization mass spectrometry verified our findings.

## Conclusion

The described methodology allows for rapid assessment of infection-specific changes in the plant metabolism, in particular of phenolics, alkaloids, oxylipins, and carbohydrates. Moreover, potential novel biomarkers can be detected and used to predict the quality of plant infections.

# BACKGROUND

Metabolism consists of a complex network of biosynthetic pathways and comprises a series of biochemical reactions that are catalyzed by enzymes [1,2]. Plants and animals produce a remarkably diverse array of over 100,000 secondary metabolites. In contrast to the primary metabolites, secondary metabolites are normally not directly involved in growth, development, and reproduction, but partly play essential roles in the adaptation of the organisms to their environments [1-5]. The rich diversity of secondary metabolites results from a co-evolutionary process with biotic attackers. Secondary metabolites stored in various plant organs and cell compartments play important roles as*constitutive defence* against herbivores and microbial pathogens. Natural selection favors those plants that have developed distinct cellular defence responses, specifically induced and tailored against the invasion of pathogens or attack by insects [1-5]. *Induced defences* against various attackers are differentially orchestrated by a network of interacting signaling pathways.

Depending on the signaling pathway employed, the defence response may in some cases only encompass the infected tissue (*local response*) or in addition involve non-infected parts of the plant (*systemic response*), thus protecting the plant against subsequent attacks [6,7]. The phytohormone salicylic acid (SA) is mainly induced by biotrophic and hemi-biotrophic parasites and initiates and controls both the systemic response and the hypersensitive response (HR) [5,8-10]. The HR is a type of programmed cell death in which the plant sacrifices a few cells immediately surrounding the attacked cell, thereby restricting pathogen growth [8-10]. For the proper development of a HR, the SA-signaling pathway includes feedback loops which integrate other defence-associated signals, such as (i) jasmonic acid (JA) [11,12], (ii) ethylene (ET) [13], and (iii) nitric oxide (NO) [14-16]. Other metabolic events, which are important in the execution of HR include reinforcement of the cell wall[17], phytoalexin production [18], expression of pathogenesis-related proteins [19], reactive oxygen species (ROS) generation [17,20], and the readjustment of the primary metabolism, that includes sugar accumulation and signaling [17,20]. The overall signature of the complex metabolic and signaling networks in the plant determines the pattern of defence-related metabolome changes and, hence, eventually the outcome of the plant - pathogen interaction [1,8,9,12].

Different methods have been developed for the analysis of multiple metabolites during the plant defence response, such as fluorescence microscopy [21-24] and mass spectrometry (MS) [25-30] Because of the high complexity of the plant metabolome, hyphenated MS-based methods are mostly preferred, *e.g.* the coupling of a mass spectrometer with gas chromatography (GC) [25-27] or liquid chromatography (LC) [28-30]. By adding reference compounds, these techniques allow the absolute quantification of individual metabolites [27-30]; however, the employment of these hyphenated methods requires cumbersome sample preparation processes [25-30].

Matrix-assisted laser desorption ionization mass spectrometry (MALDI MS) has been increasingly employed in metabolomic studies. Advantages of this analytical method are that first limited effort is required for sample preparation and second the method lends itself to high throughput analysis [31-35]. A few reports have described the utilization of ultraviolet (UV) and infrared (IR) lasers for the analysis of plant metabolites directly from tissue [32,36-40]. This is possible because laser energy in the UV or IR region can be absorbed directly by different biomolecules in the tissue. If the concentration of absorbing biomolecules is high enough, they can function as an endogenous matrix for the MALDI process. IR lasers with an emission wavelength around 3 μm, such as an Er:YAG laser emitting at 2.94 μm or an optical parametric oscillator (OPO) laser, make use of O-H containing biomolecules (*i.e.* mostly water molecules) as an endogenous matrix via excitation of the O-H bond stretch vibrations [41]. In the following, we will use the abbreviation "IR-LDI" instead of "IR-MALDI" to account for the fact that only endogenous biomolecules are used as energy absorbers.

While direct tissue analysis by mass spectrometry generally does not allow for absolute quantification of molecular concentrations, a semi-quantitative analysis may be possible by recording the changes in signal intensities of individual compounds, for instance in response to a treatment. A particularly elegant method to retrieve information about even small changes of the molecular profiles in a complex system (a LDI mass spectrum may contain hundreds of ion signals) is the use of principal component analysis (PCA) [42,43]. PCA is a multivariate data analysis which can be used to reduce the complexity of large amounts of detectable ion signals to a small number of independent (orthogonal) component vectors. Hence, in the PCA the individual identity as well as the absolute intensity of an ion signal are not as relevant as the mass spectral profile as a whole [42,43].

With respect to the mass analyzer, for the analysis of rough and electrically non-conducting surfaces (such as leaf tissue), orthogonal time-of-flight (oTOF) mass spectrometers have been found to provide particularly beneficial properties [40,44,45]. This is because these instruments decouple the ion formation process from the time-of-flight measurement and thereby provide a high and constant mass accuracy, independent of the exact morphology of the investigated sample [40,44,45]. The combination of IR-LDI and oTOF MS also provides particularly soft desorption/ionization conditions that produce a minimum of analyte ion dissociation [44,46]. The analysis of a wide range of metabolites directly from plant tissues - e.g. lipids (phospholipids, sulfates, triglycerides), carbohydrates (oligosaccharides), glycosidic flavonoids, and alkaloids - has been reported using this type of instrument [40].

Here, we used direct IR-LDI-oTOF mass spectrometry to simultaneously profile a large number of metabolites of the tobacco plant *Nicotiana tabacum* (Samsun, *SNN*), generated during development (1 to 9 hours post infection) of the HR against the oomycete *Phytophthora nicotianae*. A multivariate analysis - principal component analysis (PCA) - was used to reduce the large and complex LDI MS data to one- or two- dimensional data sets. The statistical analysis was compared with an established indicator of the biological response of the plant - pathogen system, dead cell count analysis, and an established method for studying plant defence responses, *i.e.* quantification of defence-related phytohormones using liquid chromatography in combination with electrospray ionization tandem mass spectrometry (LC-ESI MS/

MS) [28-30]. We conclude that IR-LDI MS metabolome profiling in conjunction with PCA can be a powerful tool to obtain information about plant defence cascades and to monitor the intensity and progression of the infection processes.

# RESULTS AND DISCUSSION

Although extensive data exists describing the role of various secondary metabolites (such as SA) during the hypersensitive defence response in pathogen-plant interaction [5,8-12], much remains to be learned about the response as a whole.

In a classical approach, it would be possible to identify the different signal cascades that are activated in the leaf during the infection by quantifying "key" metabolites for each sample treatment conditions (*i.e.* blank, placebo, and infection). However to reliably quantify these compounds, an internal standard would be required for each metabolite of interest to compensate for the matrix effects that originate from the inherent amount of variability associated with biological samples. Thus, the quantification approach is suitable for the verification of a hypothesis but is not recommended as the first step for any plant study when not much is known about the occurring cellular processes, as it is the case here.

Relative up- or down-regulation of metabolic cascades may be followed by employing a semi-quantitative approach. Here, we applied this method and used the earliest measurement time point obtained for each sample treatment (*i.e.* 1 hpi) as reference point. This way, we can compensate for non-biological effects (*e.g.* systemic analytical errors caused by changes in the growth environment) which could influence the metabolic response. The general procedure and experimental workflow of the analysis is depicted in Figure 1; see Materials and Methods for details.

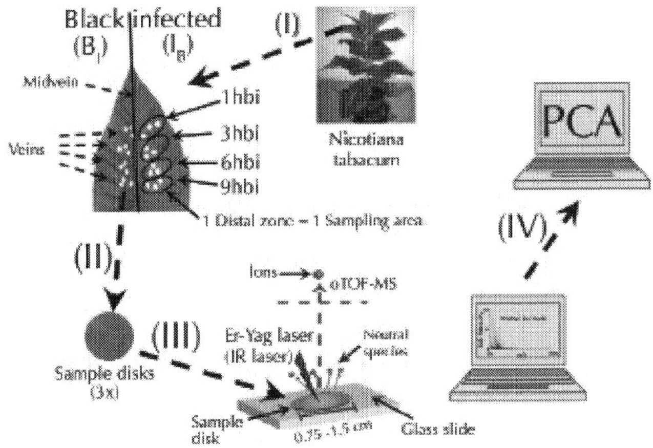


**Figure 1:** Workflow of the combined IR-LDI mass spectrometry - principal component analysis of plant metabolites. Schematic workflow of the analysis: (I) sample treatment, (II) sample collection, (III) direct IR-LDI-oTOF MS measurement, (IV) principal component analysis. The lower cuticle of the leaf is removed using adhesive tape and disks are placed upside down on a glass slide. The glass slide is mounted in a custom-made sample plate holder and transferred into the laser mass spectrometer for measurement. The nomenclature employed to identify the sample type is explained in the text.

The accessible mass range of the matrix-less LDI MS approach is limited to about 1000 *m/z* units. Because no matrix was employed in these measurements, the mass spectra are fully devoid of the matrix-derived chemical background, which can severely complicate standard MALDI MS analysis of small molecules. Moreover, signals corresponding to other relative abundant biological substances - such as peptides - also are not produced. The IR-LDI-oTOF MS measurements were performed in both positive and negative ion modes from differently treated leaf samples: zoospores infiltrated (infected, I), tap-water infiltrated (placebo, P) and untreated (blank, B), and at time points of 1, 3, 6, and 9 hpi (see Material and Methods for details on the sample preparation procedure). As an example, Additional file 1 shows the positive and negative ion mode mass spectra for the region between *m/z* 100 to 500 obtained using IR-LDI-oTOF MS for two samples, infected ($I_P$) and placebo ($P_I$), harvested at 6 hpi from a *SNN* tobacco leaf. A full list of all consistently detected ions is provided in Table 1.

Figure 2 displays the time-profiles for a subset of 9 example metabolites, as derived from the IR-LDI MS analysis of differently treated samples; data are normalized to the signal intensity obtained at 1 hpi. The chemical assignments (identities) for these 9 metabolites and several other compounds (Table 1) could be made with reasonable confidence on the basis of precise mass measurement, the isotope ratios of the corresponding signals, and metabolic databases (see Material and Methods for details on the identification of metabolites).

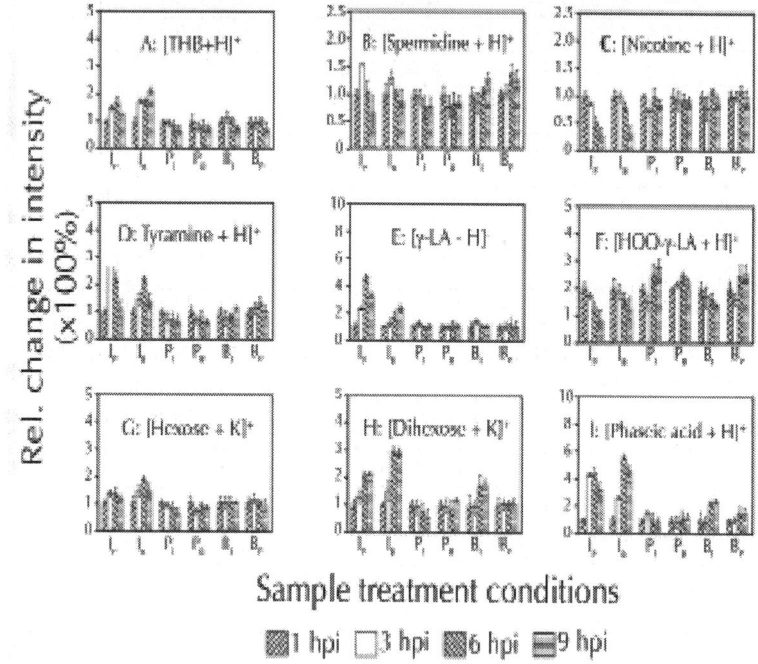


**Figure 2:** IR-LDI-oTOF MS analysis of metabolites (time-profiles). Metabolic time-profiles of selected metabolites in *Nicotiana tabacum* (cv.*SNN*) samples that were subjected to different treatments: (I) infected, (P) placebo, and (B) blank (for detailed explanation of the nomenclature employed, see text). IR-LDI MS signal intensities of the molecular ions (see Table 1 for ion type) were normalized to the value obtained at 1 hpi. THB (tetrahydrobiopterin); γ-LA (γ-linolenic acid); HOO-γ-LA (peroxidated-γ-linolenic acid).

**Table 1:** List of ion signals detected by direct IR-LDI-oTOF mass spectrometry in leaf samples of*Nicotiana tabacum* in (A) positive and (B) negative ion mode*

| **A: Positive ion mode** | | | | | |
|---|---|---|---|---|---|
| **#** | **Obs. Mass (*m/z*)** | **Chemical Formula** | **Theor. Mass (*m/z*)** | **Δm (ppm)** | **Tentative Assignment*** |
| a1 | 104.107 | $[C5H14NO]^+$ | 104.107 | 0.0 | choline |
| a2 | 122.095 | $[C8H11N + H]^+$ | 122.097 | 16.4 | phenethylamine |
| a3 | 132.112 | $[C5H13N3O + H]^+$ | 132.113 | 7.6 | N-carbamoylputrescine |
| a4 | 138.091 | $[C8H11NO + H]^+$ | 138.092 | 7.2 | tyramine |
| a5 | 146.167 | $[C7H19N3 + H]^+$ | 146.165 | 13.7 | spermidine |
| a6 | 163.122 | $[C10H14N2 + H]^+$ | 163.123 | 6.1 | nicotine or anabasine |
| a7 | 192.172 | $[C13H21N + H]^+$ | 192.175 | 15.6 | |
| a8 | 219.027 | $[C6H12O6 + K]^+$ | 219.027 | 0.0 | hexose |
| a9 | 242.126 | $[C9H15N5O3 + H]^+$ | 242.125 | 4.1 | tetrahydrobiopterin |
| a10 | 251.221 | $[C14H26N4 + H]^+$ | 251.223 | 8.0 | |
| a11 | 266.185 | $[C17H25N + H]^+$ | 266.186 | 3.8 | |
| a12 | 269.170 | $[C17H20N2O + H]^+$ | 269.165 | 15.8 | |
| a13 | 281.140 | $[C15H20O5 + H]^+$ | 281.138 | 6.0 | phaseic acid |
| a14 | 311.221 | $[C18H30O4 + H]^+$ | 311.221 | 0.0 | hydroperoxyoctadecatrienoic acid |
| a15 | 318.153 | $[C14H23NO7 + H]^+$ | 318.155 | 6.3 | glycosylated-retronecin or [tyramine + hexose][1] |
| a16 | 325.236 | $[C19H32O4 + H]^+$ | 325.237 | 3.1 | methyl-hydroperoxyoctadecatrienoate |
| a17 | 343.186 | $[C19H28O4 + Na]^+$ | 343.189 | 8.7 | |
| a18 | 355.153 | $[C19H24O5 + Na]^+$ | 355.152 | 2.8 | gibberellin 20 |
| a19 | 361.211 | $[C19H33ClO4 + H]^+$ | 361.215 | 10.0 | methyl chloro hydroperoxyoctadecatrienoate |
| a20 | 381.079 | $[C12H22O11 + K]^+$ | 381.080 | 2.6 | dihexose |
| a21 | 399.085 | $[C12H24O12 + K]^+$ | 399.091 | 15.0 | [hexose + hexose]1 |
| a22 | 431.222 | $[C23H30N2O6 + H]^+$ | 431.218 | 9.3 | |
| **B: Negative ion mode** | | | | | |
| **#** | **Obs. Mass (*m/z*)** | **Chemical Formula** | **Theor. Mass (*m/z*)** | **Δm (ppm)** | **Tentative Assignment*** |
| b1 | 103.004 | $[C3H4O4 - H]^-$ | 103.004 | 0.1 | malonic acid |
| b2 | 105.019 | $[C3H6O4 - H]^-$ | 105.019 | 2.2 | glyceric acid |

| b3 | 127.016 | [C4H4N2O3 - H]$^-$ | 127.014 | 15.7 | |
|---|---|---|---|---|---|
| b4 | 133.014 | [C4H6O5 - H]$^-$ | 133.014 | 1.3 | malic acid |
| b5 | 135.029 | [C4H8O5 - H]$^-$ | 135.029 | 1.9 | threonate |
| b6 | 146.046 | [C5H9NO4 - H]$^-$ | 146.045 | 3.0 | |
| b7 | 173.008 | [C6H6O6 - H]$^-$ | 173.009 | 3.6 | dehydroascorbic acid |
| b8 | 174.019 | [C6H7O6 - H]$^-$ | 174.017 | 11.5 | mono-dehydroascorbic acid |
| b9 | 179.055 | [C6H12O6 - H]$^-$ | 179.055 | 1.6 | hexose |
| b10 | 191.053 | [C7H12O6 - H]$^-$ | 191.056 | 16.6 | quinic acid |
| b11 | 197.062 | [C13H10O2 - H]$^-$ | 197.060 | 10.3 | |
| b12 | 215.031 | [C6H12O6 + Cl]$^-$ | 215.032 | 4.8 | hexose |
| b13 | 225.057 | [C10H10O6 - H]$^-$ | 225.061 | 17.8 | chorismic acid |
| b14 | 226.995 | [C6H8O7 + Cl]$^-$ | 226.996 | 1.8 | citric acid |
| b15 | 239.073 | [C15H12O3 - H]$^-$ | 239.071 | 8.4 | |
| b16 | 255.233 | [C16H32O2 - H]$^-$ | 255.232 | 3.0 | palmitic acid |
| b17 | 267.071 | [C9H16O9 - H]$^-$ | 267.071 | 0.3 | |
| b18 | 269.082 | [C16H1404 - H]$^-$ | 269.081 | 3.7 | |
| b19 | 277.215 | [C18H30O2 - H]$^-$ | 277.217 | 6.7 | γ-linolenic acid |
| b20 | 279.034 | [C13H12O5S - H]$^-$ | 279.033 | 1.8 | |
| b21 | 279.231 | [C18H32O2 - H]$^-$ | 279.232 | 2.1 | α-linolenic acid |
| b22 | 285.080 | [C16H14O5 - H]$^-$ | 285.076 | 14.8 | |
| b23 | 297.045 | [C16H10O6 - H]$^-$ | 297.047 | 4.4 | |
| b24 | 309.044 | [C17H10O6 - H]$^-$ | 309.040 | 12.2 | |
| b25 | 310.147 | [C19H21NO3 - H]$^-$ | 310.144 | 8.4 | |
| b26 | 313.075 | [C17H14O6 - H]$^-$ | 313.075 | 1.3 | |
| b27 | 315.090 | [C17H16O6 - H]$^-$ | 315.090 | 0.9 | |
| b28 | 325.044 | [C9H15N2O9P - H]$^-$ | 325.043 | 2.6 | |
| b29 | 327.054 | [C17H12O7 - H]$^-$ | 327.051 | 8.7 | |
| b30 | 353.088 | [C16H18O9 - H]$^-$ | 353.087 | 0.7 | chlorogenic acid |
| b31 | 359.118 | [C12H24O12 - H]$^-$ | 359.119 | 2.8 | [hexose + hexose]1 |
| b32 | 371.111 | [C20H20O7 - H]$^-$ | 371.113 | 5.8 | |
| b33 | 377.083 | [C12H22O11 + Cl]$^-$ | 377.084 | 3.9 | dihexose |
| b34 | 395.095 | [C12H24O12 + Cl]$^-$ | 395.095 | 0.8 | [hexose + hexose]1 |
| b35 | 447.132 | [C22H24O8 - H]$^-$ | 447.129 | 8.4 | |
| b36 | 457.278 | [C29H42O2 + Cl]$^-$ | 457.286 | 17.3 | |
| b37 | 477.145 | [C16H30O16 - H]$^-$ | 477.145 | 0.5 | |
| | | | | | |

## Alkaloids and Phenolics

The alkaloid *tetrahydrobiopterin* (THB, Figure 2A) was identified as a metabolite for which the concentration increases substantially exclusively during local infection. THB functions as redox-cofactor for a number of metabolic processes (e.g. the oxidation of phenylalanine). In animals, it is also a cofactor of nitric oxide synthase (NOs) [47], but the exact plant metabolism of NO, an important messenger in stress response, is still controversial [48]. *Spermidine* (Figure 2B) a polyamine involved in various cellular processes [49], exhibits a moderate transient concentration increase (at ~ 3 hpi) in infected samples, but due to its multiple functions in plant metabolism it cannot be regarded a key indicator of infection. In contrast, *nicotine* levels (and possibly also those of its isomer anabasine) significantly decrease in infected leaf samples (Figure 2C). The alkaloid nicotine, a highly effective insecticide, is produced in the tobacco roots, transported to leaves, and stored in vacuoles. Increase of nicotine concentration in tobacco leaves has been related to the JA signaling cascade in response to attack from herbivores [4,29,50]. However, in contrast to the response to herbivore attack, infection with *Phytophthora nicotianae* results in a steep decline in nicotine concentration, when compared to placebo and blank samples. This metabolic change might be correlated with an increase in SA levels, an antagonist of JA (see hormone analysis below; Figure3).

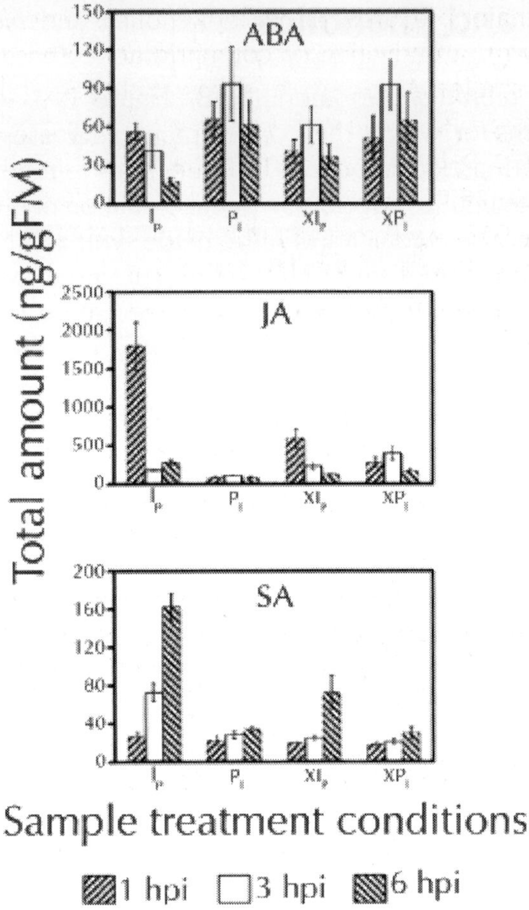


**Figure 3:** Quantitative LC-ESI MS/MS analysis of phytohormones (time-profiles). Absolute amount of the major signaling phytohormones (ABA, JA, and SA) per weight of fresh leave material in two different cultivars of Tobacco (*SNN*, and *Xanthi*), determined by LC-ESI MS/MS in multiple reaction mode (MRM). Samples labeled with a preceding "X*" represent samples from *N. tabacum* (cv. *Xanthi*), * represents the sample treatment (for detailed explanation of the nomenclature employed, see text).

Phenolic metabolites whose biosynthesis is derived from the shikimic acid pathway [1,51], exhibit an increase in concentration during infection. Examples are two decarboxylation products of phenylalanine: *tyramine* (Figure 2D) and *phenylethylamine* (data not shown); *chorismic acid* (data not shown), a product of the shikimic pathway and major precursor of plant phenolics; and*chlorogenic acid* (data not shown), a derivative of coumaric acid. These observations concur with studies reporting a substantial up-regulation of the oxidative pentose-phosphate (OPP) pathway, which creates precursors for the shikimic pathway, and the L-phenylalanine-ammonia-lyase (PAL) activity, a major control step of the phenolic metabolism [17]. *Chorismic acid* or/and cinnamic acid (the product of the PAL reaction) are potential precursors of the hormone SA [1]. SA levels strongly increase after infection with *P. nicotianae* (Figure 3; see hormone analysis below).

## Free Fatty Acids and Oxylipins

Concentrations of free fatty acids (FAs) in plant tissues, such as γ-*linolenic acid* (γ-LA, Figure 2E), *α-linolenic acid* (data not shown), and *palmitic acid* (data not shown), can be altered by wounding or infection [52-55]. For example, various oxylipins which function as stress messengers [52,54,55] are created by non-enzymatical or enzymatical (lipoxygenases) oxidation of FAs, in particular γ-LA [56-60]. Using IR-LDI-oTOF MS, three oxylipins were detected: *peroxidated γ-linolenic acid* (HOO-γ-LA, Figure 2F), *peroxidated methyl-γ-linolenate* (data not shown), and possibly *peroxidated chloride-methyl-α-linolenate* (data not shown); all are presumably derived from γ-LA [52,60]. High levels of LA›s, in particular of γ-LA (Figure 2E), are only observed in pathogen-infected samples and may reflect activation of phospholipases. For example, phospholipase A2 ($PLA_2$) is known to be activated in plants by infection [61] and constitutes the first step in the biosynthesis of the hormone JA. Increased levels of JA are indeed transiently found in infected samples at 1 hpi (see below; Figure 3). The decrease in HOO-γ-LA (Figure 2F) and JA abundances (while γ-LA increases) in locally and not-locally infected samples, which were not under mechanical-stress (such as $B_l$ samples), most likely reflects the infection - induced inhibition of pathways employed for lipid oxidation. This assumption is supported by the fact that the decrease in oxylipins and JA levels was not observed in samples that experience pure local mechanical-stress (which is known to activate the JA pathway). Moreover, the SA signaling cascade has been reported to block the first step in the lipoxygenase (LOX) cascade, which is responsible for the biosynthesis of JA from its precursor γ-LA [8,52].

## Carbohydrates

Recently, it was demonstrated for the tobacco - Phytophthora system that during development of the HR sugars (hexoses and dihexoses) accumulate in the cell wall [17,20]. Using IR-LDI oTOF MS, hexoses and dihexoses are detected in both positive and negative ionization modes. The time profiles of the potassiated sugar molecules are shown in Figure 2G & 2H, respectively. Although isomers cannot be differentiated in the MS profile, glucose and fructose are expected to comprise the major hexoses and sucrose the major dihexose in plants. In agreement with earlier biochemical analysis[17], the MS analysis indicates a significant increase in sugar levels (in particular for the dihexose) in infected samples.

## Plant Hormones

ABA, a regulator of various environmental stress responses [62-64], positively affects the early (pre-invasion) defence [65]. It also influences the effect of the volatile hormone ET (which regulates the antagonistic signaling cascades of SA and JA) [66], and inhibits late defence reactions (such as the SA-mediated HR) [67]. The levels of *phaseic acid* (Figure 2I), the first stable decomposition product of ABA [68,69], increased exclusively in the infected samples, reflecting the possible degradation of ABA (see below, Figure 3)

Because IR-LDI MS ion signals of SA, JA, and ABA were either too low in intensity (SA) or not detectable (JA, ABA), we extracted these hormones from the plant lamina and used ESI-LC MS/MS to determine their absolute concentration in tissue [28-30]. The metabolic time profiles of the three hormones are displayed in Figure 3. The figure includes data obtained from two different cultivars of*Nicotiana tabacum*, the resistant cultivar *SNN* and the nonresistant cultivar *Xanthi*, which exhibits no or only a weak HR. In *SNN*, high JA levels appear transiently during the first hour post infection [70]. The rapid decline in JA after its initial burst (which is most likely due to the inhibition of the LOX cascade of the JA biosynthesis pathway) is possibly reflected in the IR-LDI-oTOF mass spectra by the "cross-over" between γ-LA (for which signals increase) and its peroxidation products (for which signals decrease). In contrast to JA, the levels of the phenolic hormone SA, known to inhibit JA biosynthesis [8,52], increase monotonously with time after injection of zoospores. The increase in SA concentration in the tobacco - Phytophthora system after 1 hpi is further supported by the appearance of PR1, a marker for the SA pathway [17]. In the LDI MS profile, infection-induced biosynthesis of SA may be reflected by the appearance of its phenolic precursors *chorismic acid* and*chlorogenic acid* (data not shown) [1,12]. In *Xanthi*, the initial JA burst was small and the increase in SA levels was retarded, compared to those seen in *SNN*, reflecting the difference in their response against the pathogen.

## Principal Component Analysis

A PCA transformation can help to visualize the most significant differences in the mass profiles between samples and allows similar samples to be grouped together. The first principal component (PC1) is most strongly influenced by the combination of ion signals that exhibit the largest change between the recorded spectra. In a similar way, subsequent principal components (PC2, PC3, etc) represent less influential changes that can, however, still help to identify particular features of the investigated system. Figure 4 shows the PC1 *vs.* PC2 score plot of the unsupervised PCA performed with the positive mode ion spectra (see Material and Methods for details). This graphic representation reveals a high correlation between the MS profiles of the six different groups of investigated samples. Based on PC2 values, they are essentially grouped into one of three "qualitative" clusters. Qualitatively, these clusters may be seen as representing: (I) untreated samples (blank near a placebo zone, $B_P$); (II) "stressed" samples from local water infiltration (placebo near blank or infection zone; $P_B$, $P_I$); and (III) "stressed" samples from infection ($I_P$, $I_B$, $B_I$). This qualitative analysis suggests that PC2 may capture both systemic and local stress signals, though the local stress response is presumably reflected stronger. In the score plot, samples are not differentiated with respect to the time post treatment. As can be seen in Figure 5, mass spectra of the infected samples tend to show an increase of the PC1/PC2 values derived from their metabolic profile with time post infection, while the picture for the other sample treatments is less clear.

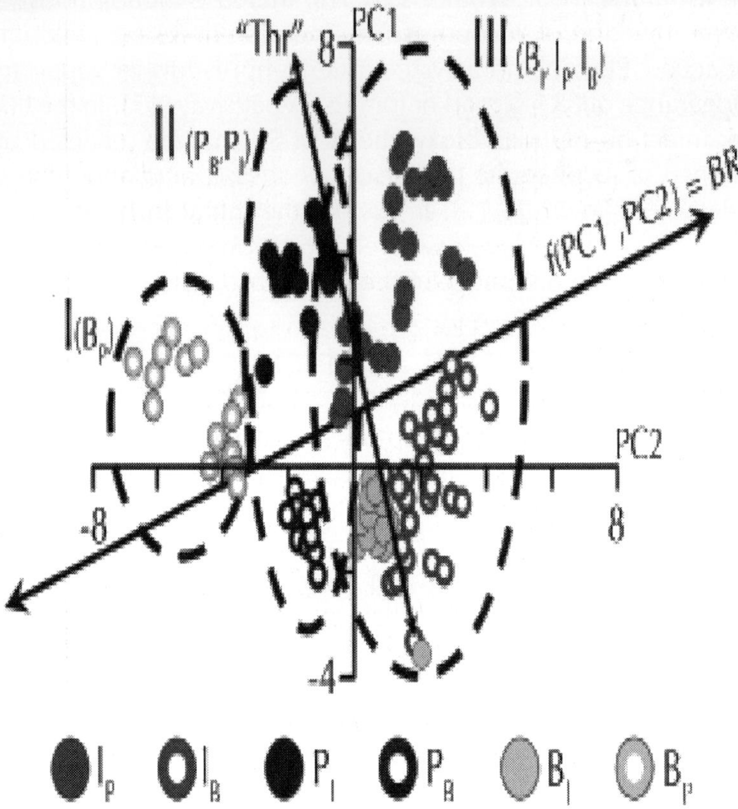


**Figure 4:** Principal component analysis of the IR-LDI-oTOF MS spectra. PC1 *vs*. PC2 score plot of an unsupervised PCA of the positive ion mode IR-LDI mass spectra acquired from differently treated samples of*Nicotiana tabacum* (cv. *SNN*). Based on their distribution with respect to the PC2-axis, the samples can, qualitatively, be grouped into three clusters: (I) untreated samples (blank near a placebo zone, $B_P$); (II) "stressed" samples from local water infiltration (placebo near blank or infection zone; $P_B$, $P_I$); and (III) "stressed" samples from infection ($I_P$, $I_B$, $B_I$). The line labeled as "Thr" separates samples that do not undergo HR (-HR) from those that undergo HR (+HR) (*i.e.* top right = high cell death, bottom left = low cell death). The line f(PC1,PC2) is displayed as the perpendicular line to "Thr". The "stress status" (BRI value, displayed in Figure 5) for each sample is obtained from the projection of any point to f(PC1,PC2).

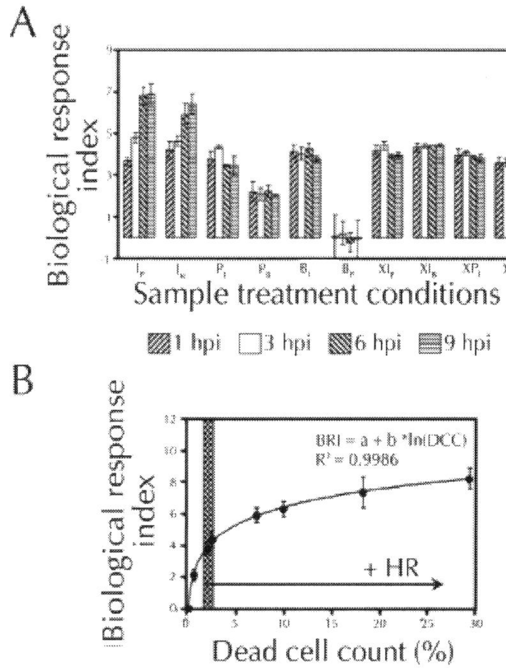


**Figure 5:** Biological response index and comparison with dead cell count values. (A) Biological response index (BRI, f(PC1,PC2)) obtained from the PCA shown in Figure 4 (see text for detailed explanation of the procedure). Samples labeled with a preceding "X*" represent samples from*N. tabacum* (cv. *Xanthi*), * represents the sample treatment (see text for details). (B) Comparison between BRI and the dead cell count (DCC, %) determined for differently treated *SNN* samples. Solid line: Best fit according to BRI = a + b*Ln(DCC), where a = 2.6044 and b = 1.6535 ($R^2$ = 0.9986). The shaded area indicates the threshold in terms of DCC that defines a positive HR.

## Identification of Key Metabolites and Their Pathways

The corresponding loading plots (displayed in Additional file 2) identify which metabolites, and therefore their associated pathways, contribute the most to the variance of the two principal components. Some of these metabolites can directly be related to known metabolic processes. Examples for PC1: *phaseic acid* (*m/z* 281.140, *cf.* to Table

1 for the type of ion formation) relates to ABA turnover (see hormone analysis above); *tyramine* (*m/z* 138.091, *cf.* to Table 1) relates to the phenolic metabolism and SA turnover (see hormone analysis above). Examples for PC2: HOO-γ-LA (*m/z* 311.220, *cf.* to Table 1) and its methyl ester (*m/z* 325.236, *cf.* to Table 1) relate to LOX activity and JA turnover (see hormone analysis above). The loading plots also revealed MS signals that exhibit a high correlation in the PCA, for which the identity of the corresponding ions, however, is currently unknown. For example, the ion signal with an *m/z* value of 318.153 (Table 1) shows a strong correlation with PC1. Based on its mass, this metabolite could possibly be either a glycosylated ester of the alkaloid *retronecin* [71], or a gas phase adduct of protonated *tyramine* + hexose (or both). A second signal with an *m/z* value of 361.211 (Table 1) shows a strong correlation with PC2. This species might be peroxidated-chlorinated-α-methyl-linolenate or a gas phase adduct of protonated*peroxidated-methyl-γ-linolenate* with hydrochloric acid. Further studies are needed to elucidate the chemical identity of these species. These examples demonstrate that the defence response of tobacco (*SNN*) toward *Phytophthora Nicotianae* can only be more fully captured by the combination of the two main principal components (PC1 & PC2).

## Biological Response Index

Because only locally infected tissue of the *SNN* cultivar undergoes HR, while neighboring leaf areas do not exhibit HR, two qualitatively lines can be drawn in the PC1 *vs.* PC2 score plot of Figure 4. The first one, labeled as "Thr" (threshold), is a line that separates all measured samples from tobacco (*SNN*) into two regions: (a) +HR (undergoing HR) and (b) -HR (not undergoing HR), respectively. The second one is perpendicular to "Thr", and is labeled as f(PC1,PC2). The projection of the PC1 *vs.* PC2 data points onto this line - f(PC1,PC2) - can be regarded as reflecting the "stress status" of the samples. One may thus call this one-dimensional projection an integrated "Biological Response Index" (BRI).

Figure 5A shows the so-derived BRI values for the differently treated *SNN* and *Xanthi* samples for time points between 1 to 9 hpi. For convenience, the mean of the BRI values derived for "untreated" samples ($B_P$), where a minimum of metabolic effects can be expected to occur, was furthermore arbitrarily set to zero. The highest BRI values of ~7 were derived for infected *SNN* samples at 6 and 9 hpi, while infected *Xanthi* samples exhibit intermediate values of ~4, which are essentially constant at all time points, and also similar to the BRI values obtained for $P_I$ and $B_I$ *SNN* samples.

To further evaluate the possible relevance of the BRI, we examined how well the BRI values correlate with the number of cells that had undergone HR by performing classical dead cell count analysis of identically treated samples. The correlation between the two values is plotted in Figure 5B. Cell death is a consequence of (SA - mediated) HR and proportional to the strength of the plant defence response. The BRI *vs*. DCC plot (Figure 5B) reveals a non-linear function of the form "a + b*Ln(DCC)" between the qualitative BRI and the quantitative DCC values. The non-linearity of the relationship may be interpreted as describing the complex metabolic (and hormonal) transition, starting from initial local or "micro-systemic" stress or infection responses to a specific defence reaction. BRI values around 5 reflect the transition from more or less reversible stress response to irreversible processes (such as HR) which inevitably result in cell death (represented by the line labeled "Thr" in Figure 4), but this transition is not abrupt.

The BRI parameter appears to only reflect the strength of the local defence response toward the infection. Treatments applied on the opposite side of the midvein did not notably influence the BRI value obtained for the infected samples. This demonstrates that at least within the time frame of 1 - 9 hpi, the analysis of leaf sections which develop a HR will not be dramatically affected by major systemic effects which traversed the major veins (Additional file 3). However, the occurrence of more minor systemic effects, that are influencing the overall metabolic profile, is well reflected by the original PCA score plot. For example, in the PC1 *vs*. PC2 score plot of Figure 4, the metabolic profile of blank control samples is placed adjacent to an infection site, suggesting that PC2 may capture "micro-systemic" metabolic responses.

Employing loading plots (Additional file 2) to find those metabolites that are associated with a high BRI value, two ion species with *m/z* values of 269.170 and 431.222, respectively, were identified to exhibit a particular high correlation with the BRI. The metabolic time profiles of these, as yet unidentified compounds, are displayed in Figure 6. Both species essentially show the same expression profile with time post treatment. The mass difference of 162.052 u suggests that these two metabolites likely represent the deglycosylated/glycosylated derivative of each other, where the sugar moiety would be a hexose. The strong correlation with the BRI value and thus to the infection-related stress response renders them as putative biomarkers for monitoring the HR defence response. Future studies will aim to structurally characterize the two compounds.

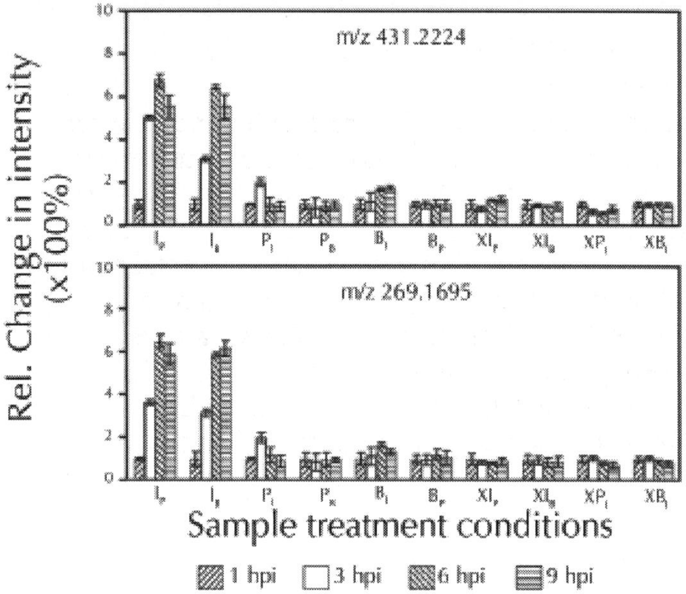


**Figure 6:** Putative biomarkers for the HR of tobacco (*SNN*). Metabolic time-profiles (determined by direct IR-LDI-oTOF MS) of two, as yet unidentified metabolites exhibiting *m/z* values of 431.222 and 269.170 that exhibit a strong correlation with the biological response index (BRI, f(PC1,PC2)) and could be employed for monitoring the defence response of*N. tabacum* (cv. *SNN*) toward *P. nicotianae* infection.

# CONCLUSIONS

IR-LDI-oTOF mass spectrometry allowed the rapid and simultaneous detection of a wide range of naturally occurring (unlabeled) metabolites directly from leaf lamina samples of tobacco plants. Changes in time for the signals of secondary metabolites (such as phenolics, alkaloids, and oxylipins) and primary metabolites (such as hexoses and dihexoses) were recorded for differently treated samples. The recorded changes in the metabolome indicated that the JA/SA-defence cascades as well as the ABA turnover cascade were involved in stress/infection response. The involvement of these three cascades can explain the majority of the physiological changes that were previously observed during this plant - pathogen infection [17,20]. The hypothesis that the JA/SA-defence cascades as well as ABA turnover are important for the development of HR was successfully verified by LC-ESI-MS/MS analysis of phytohormones.

From the PCA of the positive ion mode spectra a qualitative stress response factor (denominated: biological response index, BRI) was derived. By employing the PCA loading plots, several ion signals were identified to be strongly correlated with the local defence response against the pathogen. While many of these metabolites have been described previously to be involved in stress/infection response, two ion signals that are showing a particular strong correlation with the BRI could so far not be assigned. Potentially, the two compounds can serve as biomarkers for monitoring the HR. Thus, IR-LDI-oTOF MS analysis in combination with PCA may help to not only to rapidly record complex metabolic transitions during specific stress responses, but also to identify such potential biomarkers.

Furthermore, two additional advantages of the LDI-oTOF MS method are that it first allows to record metabolic profiles directly from fresh tissue and second provides a high lateral resolution. This simplifies and accelerates the analysis, on the one hand, and facilitates the mapping of the spatial spread of the metabolic response across the leaf on the other. Recent developments even indicate the possibility to analyze individual plant cells [39]. Thus, in nearby future; this method could be beneficial for studying developmental processes, which exhibit distinctive spatial orientation as well as several types of stress responses. In the case of the defence of the *N. tabacum* cultivar *SNN* toward*P. nicotianae*, the HR is strictly a localized process, however the defence response may not need to be (there might be signals going out which prepare neighboring cells to execute HR faster). Hence, future high resolution metabolome mapping could be an extremely powerful tool, e.g. for detecting the spread of an infection and for analyzing the interaction between infected and non-infected cells.

# MATERIALS AND METHODS

## Chemicals

Deuterated abscisic acid ($D_2$-ABA) and jasmonic acid ($D_2$-dihydro-JA) were purchased from OlChemIm Ltd (Olomouc, Czech Republic); deuterated salicylic acid ($D_4$-SA), and all other chemicals/solvents were purchased from Sigma-Aldrich (Steinheim, Germany) and were used without further purification.

## Plant and Pathogen Material

Plants were grown in a climate chamber with 24°C and 22°C (day and night temperature, respectively), with a 14 hour photoperiod (light levels of 300 µmol quanta·$m^{-2}$·$s^{-1}$). Source leaves of six to eight weeks old plants from two cultivars of *Nicotiana tabacum* were used. Because only about two source leaves of this age are found in an individual tobacco plant, several identically treated plants were employed for this study. The cultivar *SNN* is resistant against the employed pathogen and capable of developing a strong HR; the other cultivar, *Xanthi*, is susceptible and does not produce sizable HR (thus, it was employed as a biological reference) [17]. Cultivation of *Phytophthora nicotianae* van Breda de Haan isolate 1828 (DSMZ, Braunschweig, Germany), the production of zoospores, and the infection procedure were performed as described previously [17,20].

## Preparation of Samples for IR-LDI-oTOF MS

Three types of samples were collected from plant leaves: Blank samples (B) were from untreated leave areas, placebo samples (P) from areas which got sterile tap water injected to represent the pure injection stress, and infected samples (I) which where injected a suspension of 750 ± 250 zoospores per µl. In case of the P and I samples, several injections of either tap water or zoospore suspension were placed close to each other within an area of ~20 $cm^2$; the total injected volume per sample was 1 ml. The liquid was slowly infiltrated with a fine syringe to fill the air space in the *spongy mesophyll*of the leaves. The benefit of employing *Phytophthora nicotianae* (which initiates local HR responses in leaves of resistant tobacco) as a pathogen and using an artificial infection procedure (local infiltration of leaves) is that the infection area is well defined and the mesophyll cells respond nearly synchronously. To determine whether, besides the local reactions of interest, there are also systemic effects, combinations of the three treatments were applied to three sets of leaves. One specific treatment was applied on one side of the midvein and the second was applied to the other. Subscripts identify this treatment; for example, $I_p$ would be a sample taken from a zoospore-injected site with a placebo (*i.e.* water injection) on the opposite side of the midvein.

For each treatment, injections were placed on a given side of the leaf in areas, where the different time points were separated by side veins (*i.e.* one distal zone for each time point) in order to minimize crosstalk between different samples [72]. Typically, three discs of 7.5 - 15 mm in diameter were cut out of each treatment area and analyzed by IR LDI-oTOF MS. The punch sites were all placed as close as possible to the point of injection. A sketch illustrating the example of a leave with spore injection on the right side of the leaf and a blank of the left is shown in Figure 1.

Samples were shock-frozen in liquid nitrogen immediately after dissection, and subsequently stored at -80°C until the MS measurements. Immediately prior to the IR-LDI MS measurement, and while the leaf was still frozen, the lower cuticle of the leaves was removed using adhesive tape. Leaf samples were placed on a standard microscope glass slide "upside down" (*i.e.* the lower epidermis side of the leaf was "face-up") using adhesive (G304, Plano W. Plannet GmbH, Wetzlar, Germany) (Figure 1). The glass slide was mounted into a custom-made MALDI sample plate using double-sided tape.

To visualize the changes in the metabolic profile that are specifically induced by infection from those initiated by mechanical or osmotic stress (due to leaf infiltration or wounding by sample collection) IR-LDI mass spectra acquired from pathogen-infiltrated (infected) leaf sections were compared with those recorded from water-infiltrated (placebo), and untreated (blank) leaf sections. Furthermore, to minimize the local variations in the amount of zoospores present in the tissue, mass profiles were taken from the middle of the injection zones. A sketch illustrating this procedure is provided in Additional file 4.

Samples were collected at time points of 1, 3, 6, and 9 hpi. Typically, the injected water exuded/evaporated within 45 min. Because a low mass spectral quality was obtained from plant tissue containing excessive water in the air space of the spongy mesophyll,1 hpi was chosen as the earliest measurement time point. By comparing changes of the relative intensity of an ion signal determined at the different time points, normalized to its intensity at the reference point of 1hpi, "key metabolites" that exhibit significant changes with time post treatment can be identified. Care was taken that samples did not reside inside the fine vacuum of the MS ion source longer than 10 min to avoid excessive water loss by dehydration.

## IR-LDI-oTOF MS Analysis of Metabolites

The mass spectrometer has been described in detail before [34,40]. This instrument is equipped with an UV-laser ($N_2$ laser, $\lambda$ = 337 nm) and an Er:YAG laser (Speser, Spektrum, Berlin, Germany). The latter is an IR-laser that emits pulses of ~150 ns duration at a wavelength of 2.94 μm and at a repetition rate of 2 Hz. The IR-laser beam irradiated the leaf surfaces at an angle of incidence of 45°. The focal spot size was slightly elliptical and had an area of ~150 × 200 $\mu m^2$ ($1/e^2$ intensity definition), the beam profile was near-Gaussian. $N_2$ was used as buffer gas for collisional cooling and focusing [44]. The pressure in the sample region of the ion source was adjusted to ~0.7 mbar [46]. For ion generation, a laser fluence (energy per pulse and area) of ~25 $kJ \cdot m^{-2}$ was typically applied. At this fluence, tissue layers of about 50 μm in depth are removed per laser pulse in the central part of the focused beam, allowing the analysis of the content of lysed cells. 120 laser shots where accumulated for each single spectrum. These successive exposures were applied over an area of typically 2 $mm^2$ by manually moving the sample target under the fixed laser beam in a zigzag pattern. This way, any given sample location was exposed to no more than four laser pulses, thereby avoiding a full penetration of the sample. Two mass spectra were recorded from each individual sample disk.

Measurements were performed in positive and negative ion mode using time-of-flight voltages of ± 10 kV. The mass resolution (FWHM) was about 6,000 in both ion modes. Using the monoisotopic ion signals of Angiotensin I ($[M+H]^+$ *m/z* 1296.6853) and 2,5-dihydroxybenzoic acid (DHB) UV-MALDI matrix molecules ($[M+H]^+$ *m/z* 155.0344) for external calibration, a mass accuracy of about ~20 ppm was achieved for the *m/z* range between 100 - 500. All ions were detected as singly charged species. With internal calibration, using signals of metabolites of known identity (e.g. nicotine, hexose, and dihexose), the mass accuracy could be improved to ~6 ppm. Mass spectra were partially processed using MoverZ software (freeware edition, Genomics Solution, Ann Arbor, MI, USA).

Although HR signals are locally triggered in the pathogen infection area [17,21], plants possess the ability to generate disease resistance against further infections in healthy (uncompromised) parts by employing secondary metabolites as infection messengers. As shown in Additional file 4, the location from where the sample disk was taken within the same distal zone may to some extent influence the metabolic pattern in the mass spectra, due to changes in the local concentrations of zoospores in the tissue. To account for this source of variability, all sample disks were taken from the center of the placebo or infected areas located inside one distal zone and as equidistance as possible to the multiple points of injection. In the case of the blank samples, the sample disks were taken from the center of their respective distal zones (Additional file 4).

If the samples were taken from the center of their respective distal zones, the main parameters that influenced the relative signal intensities and ion patterns were the amount of water contained in the leaves and the applied laser fluence (see above, preparation of samples for IR-LDI-oTOF MS). Attempts to improve a quantitative data evaluation by finding an endogenous "internal standard" were not successful. Therefore, care was taken to follow a consistent preparation and measurement protocol. Under these conditions, the variation in the intensities of specific metabolite signals obtained from identically prepared and treated samples (from the same or different plants) was between 5 - 30%. This is illustrated by the error bars in the graphs.

## Identification of Metabolites

Table 1 lists the presumable chemical composition and type of ion formation for all ~60 consistently detected ion species, in positive (Table 1A) and negative (Table 1B) ion mode, along with the experimental and calculated *m/z* values. Positively charged ions are mostly detected as protonated $[M+H]^+$ and partially as sodiated/potassiated $[M+Na]^+$/$[M+K]^+$ species, negatively charged ions as deprotonated $[M-H]^-$ and partially as chlorinated molecules $[M+Cl]^-$ (identified by the distinct $^{35}Cl/^{37}Cl$ isotopic distribution of chlorine).

In addition, a few gas phase adducts such as $[hexose + hexose + K]^+$ and $[tyramine + hexose + H]^+$ are presumably also detected due to the soft desorption/ionization conditions [40]. Since the oTOF instrument does not provide a tandem MS option for structural characterization of ions, isobaric species can not be differentiated. Hence, putative identification of metabolites is based on the comparison of experimental and theoretical mass values (which had to coincide within 20 ppm), the comparison of experimental and theoretical isotopic patterns, and the use of various databases addressing general plant metabolites, as well as literature addressing *Nicotiana tabacum* metabolism.

## Principal Component Analysis

The oTOF mass spectra were converted to ASCII files using the custom-made data acquisition software of the mass spectrometer (TOFMA v99.6; developed by W. Ens and colleagues at the University of Manitoba, Winnipeg, CA). The ASCII files were converted into Microsoft excel sheets using a home-made program to become compatible with the PCA program RapidMiner 4.600 (Rapid-I, Dortmund, Germany). Only ion signals that were consistently detected with a signal-to-noise (*s/n*) ratio of at least 10 at any of the monitored time points for *SNN* plants were used for the unsupervised PCA. Because the focus of this study was on the resistant *SNN* cultivar, those ions which were unique to *Xanthi* and not seen in *SNN* with the necessary *s/n* ratio were excluded from the PCA.

## Preparation of Samples for LC-ESI MS

LC-ESI MS quantitation of selected metabolites (phytohormones) was performed based on the work published by the Baldwin group (Max Planck Institute for Chemical Ecology in Jena, Germany) [28-30]. The infection procedure was identical to the one employed for the IR-LDI-oTOF MS measurements. Eight disks of 0.9 mm in diameter were cut out with a puncher. Samples were shock-frozen in liquid nitrogen immediately after dissection and subsequently stored at -80°C in 2 ml vials. Free phytohormones were extracted by homogenizing ~100 mg of leaf material in a SamplePrep 2000 Geno/Grinder (SPEX CertiPrep, Metuchen, NJ, USA), employing FastPrep tubes containing 900 mg of lysing matrix (BIO 101; Vista, California, USA) and 1 ml of ethyl acetate spiked with 100 ng of deuterated internal standards $D_4$-SA and $D_2$-ABA; and 40 ng of $D_2$-dihydro-JA. Phytohormones were extracted by reciprocal shaking at 6.5 m $s^{-1}$ for 2 min; consecutively, they were centrifuged at 13,000 rpm for 20 min at 4°C. The supernatants were evaporated to dryness in a vacuum concentrator at 30°C. The dry residue was re-dissolved in 600 µl of methanol (70%) and centrifuged at 13,000 rpm for 10 min at 4°C. An aliquot (500 µl) was transferred to HPLC vials for measurement.

## LC-ESI MS/MS Analysis of Phytohormones

The transferred aliquots were quantified using an LC-ESI tandem mass spectrometry system (Varian 1200, Palo Alto, USA). 10 µl of each extracted sample were injected onto a ProntoSIL column (C18; 5 µm, 50 × 2 mm; Bischoff, Leonberg, Germany) attached to a precolumn (C18, 4 × 2 mm; Phenomenex, Aschaffenburg, Germany). The mobile phase comprised solvent A (0.05% formic acid) and solvent B (methanol) used in a gradient mode (time/concentration [min/%] for B: 0.0/15; 1.5›/15; 4.5›/98; 12.5›/98; 13.0›/15; 15.0›/15) with a flow of (time/flow [ml/min]: 0.0›/0.4; 1.0›/0.4; 1.5›/0.2; 10.0›/0.2; 10.5›/0.4; 15.0›/0.4). Compounds were detected as negative ions in multiple reaction monitoring (MRM) mode. Molecular ions $[M-H]^-$ at *m/z* 137, 209, 263 (SA, JA, and ABA, respectively) and 141, 211; 265 (deuterated derivatives of the compounds, respectively) were fragmented using 15 eV collision energy. The ratios of ion intensities of non-deuterated and deuterated product ions were used to quantify endogenous SA, JA, and ABA.

## Dead Cell Count Analysis

The infection procedure was identical to the one employed for the IR-LDI-oTOF MS measurements. Dead cell count was performed with 0.5 mg $ml^{-1}$ propidium iodide aqueous solution based on the work published by the Weis group (Institute of Botany, Westfälische Wilhelms-Universität Münster, Germany) [21]. Cells with disrupted membranes (dead cells) allow the fluorescence dye propidium iodide to enter into the cytoplasmatic region, thus indicating cell death. The fluorescence was detected at 590 - 650 nm after excitation at 488 nm using a confocal laser scanning microscope (TCS SPE2 with inverse DM IRB-microscope, Leica, Wetzlar, Germany) [21].

# ACKNOWLEDGEMENTS

We are grateful to Stefan Berkenkamp and Sequenom GmbH for providing use of their oTOF instrument, Matthias Schöttner for his support with the LC-ESI-MS/MS phytohormone measurements, Jens Soltwisch for his support with the IR-LDI-oTOF MS measurements, Ina Schmitz-Thom for her support with the dead cell count measurements, Dr. Jutta Essmann and Dr. Aleš Svatoš for fruitful discussions and support of this project. AJI gratefully acknowledges financial support of the Alexander von Humboldt foundation.

# REFERENCES

1. Dixon RA: Progress natural products and plant disease resistance. *Nature* 2001, 411:843-847.
2. Nugroho LH, Verpoorte R: Secondary metabolism in tobacco. *Plant Cell Tiss Org* 2002, 68:105-125.
3. Baldwin IT: An Ecologically motivated analysis of plant-herbivore interactions in native tobacco. *Plant Physiol* 2001, 127:1449-1458.
4. Baldwin IT, Schmelz EA: Constraints on an induced defence: the role of leaf area. *Oecologia* 1994, 97:424-430.
5. Halim VA, Eschen-Lippold L, Altmann S, Birschwilks M, Scheel D, Rosahl S: Salicylic acid is important for basal defence of solanum tuberosum against phytophthora infestans. *Mol Plant Microbe In* 2007, 20:1346-1352.
6. Eyendi A, Yalpani N, Silverman P, Raskin I: Signal molecules in systemic plant resistance to pathogens and pests. *Cell* 1992, 70:879-886.

7. Pieterse CMJ, van Wees SCM, van Pelt JA, Knoester M, Laan R, Gerrits H, Weisbeek PJ, van Loon LC: A novel signaling pathway controlling induced systemic resistance in arabidopsis. *Plant Cell* 1998, 10:1571-1580.

8. Yuan S, Lin H-H: Role of salicylic acid in plant abiotic stress. *Z Naturforsch C* 2008, 63:313-320.

9. Alvarez ME: Salicylic acid in the machinery of hypersensitive cell death and disease resistance. *Plant Mol Biol* 2000, 44:429-442.

10. Vorwerk S, Schiff C, Santamaria M, Koh S, Nishimura M, Vogel J, Somerville C, Somerville S:EDR2 negatively regulates salicylic acid-based defences and cell death during powdery mildew infections of Arabidopsis thaliana. *BMC Plant Biol* 2007, 7:35-48.

11. Spoel SH, Koornneef A, Claessens SMC, Korzelius JP, van Pelt JA, Mueller MJ, Buchala AJ, Métraux J-P, Brown R, Kazan K, van Loon LC, Dong X, Pieterse CMJ: NPR1 modulates cross-talk between salicylate- and jasmonate-dependent defence pathways through a novel function in the cytosol. *Plant Cell* 2003, 15:760-770.

12. Shah J: The salicylic acid loop in plant defence. *Curr Opin Plant Biol* 2003, 6:365-371.

13. Amirsadeghi S, Robson CA, McDonald AE, Vanlerberghe GC: Changes in plant mitochondrial electron transport alter cellular levels of reactive oxygen species and susceptibility to cell death signaling molecules. *Plant Cell Physiol* 2006, 47:1509-1519.

14. Zottini M, Costa A, De Michele R, Ruzzene M, Carimi F, Lo Schiavo F: Salicylic acid activates nitric oxide synthesis in Arabidopsis. *J Exp Biol* 2007, 58:1397-1405.

15. Gould KS, Lamotte O, Klinguer A, Pugin A, Wendehenne D: Nitric oxide production in tobacco leaf cells: a generalized stress response? *Plant Cell Environm* 2003, 26:1851-1862.

16. Planchet E, Sonoda M, Zeier J, Keiser WM: Nitric oxide (NO) as an intermediate in the cryptogein-induced hypersensitive response - a critical re-appraisal. *Plant Cell Environm* 2006, 29:59-69.

17. Scharte J, Schön H, Weis E: Photosynthesis and carbohydrate metabolism in tobacco leaves during an incompatible interaction with Phytophthora ncotianae. *Plant Cell Environm* 2005, 28:1421-1435.

18. Hammerschmidt R: Phytoalexins: what have we learned after 60 years? *Plant J* 1999, 37:285-306.
19. Van Loon LC, Van Strien EA: The families of pathogenesis-related proteins, their activities, and comparative analysis of PR-1 type proteins. *Physiol Mol Plant P* 1999, 55:85-97.
20. Essmann J, Schmitz-Thom I, Schön H, Sonnewald S, Weis E, Scharte J: RNA Interference-Mediated Repression of Cell Wall Invertase Impairs Defence in Source Leaves of Tobacco. *Plant Physiol* 2008, 147:1288-1299.
21. Huang WE, Huang L, Preston GM, Naylor M, Carr JP, Yanhong L, Singer AC, Whiteley AS, Wang H: Quantitative in situ assay of salicylic acid in tobacco leaves using a genetically modified biosensor strain of Acinetobacter sp. ADP1. *Plant J* 2006, 46:1073-1083.
22. Siebka K, Weis E: Assimilation images of leaves of Glechoma hederacea: analysis of non-synchronous stomata related oscillations. *Planta* 1995, 196:155-165.
23. Yu P, Block H, Niu Z, Doiron K: Rapid characterization of molecular chemistry, nutrient make-up and microlocation of internal seed tissue. *J Synchrotron Radiat* 2007, 14:382-390.
24. Wardman P: Fluorescent and luminescent probes for measurement of oxidative and nitrosative species in cells and tissues: Progress, pitfalls, and prospects. *Free Radical Bio Med* 2007, 43:995-1022.
25. Schmelz EA, Engelberth J, Alborn HT, O'Donnell P, Sammons M, Toshima H, Tumlinson JH:Simultaneous analysis of phytohormones, phytotoxins, and volatile organic compounds in plants. *P Natl Acad Sci USA* 2003, 100:10552-10557.
26. Lytovchenko A, Beleggia R, Schauer N, Isaacson T, Leuendorf JE, Hellmann H, Rose JKC, Fernie AR: Application of GC-MS for the detection of lipophilic compounds in diverse plant tissues. *Plant Methods* 2009, 5:4-14.
27. Gaquerel E, Weinhold A, Baldwin IT: Molecular interactions between the specialist herbivore Manduca sexta (Lepidoptera, Sphigidae) and its natural host Nicotiana attenuata. VIII. An unbiased GCxGC-ToFMS analysis of the plant's elicited volatile emissions. *Plant Physiol* 2009, 149:1408-1423.

28. Kallenbach M, Baldwin IT, Bonaventure G: A rapid and sensitive method for the simultaneous analysis of aliphatic and polar molecules containing free carboxyl groups in plant extracts by LC-MS/MS. *Plant Methods* 2009, 5:17-27.

29. Keinänen M, Oldham NJ, Baldwin IT: Rapid HPLC screening of jasmonate-induced increases in tobacco alkaloids, phenolics, and diterpene glycosides in Nicotiana attenuata. *J Agric Food Chem* 2001, 49:3553-3558.

30. Schmidt L, Hummel GM, Schöttner M, Schurr U, Walter A: Jasmonic acid does not mediate root growth responses to wounding in Arabidopsis thaliana. *Plant Cell Environmen* 2010, 33:104-116.

31. Kaletaş BK, van der Weil IM, Stauber J, Dekker LJ, Güzel C, Kros JM, Luider TM, Heeren RMA:Sample preparation issues for tissue imaging by imaging MS. *Proteomics* 2009, 9:2622-2633.

32. Fraser PD, Enfissi EMA, Goodfellow M, Eguchi T, Bramley PM: Metabolite profiling of plant carotenoids using the matrix-assisted laser desorption ionization time-of-flight mass spectrometry. *Plant J* 2007, 49:552-564.

33. Gholipour Y, Nonami H, Erra-Balsells R: Application of pressure probe and UV-MALDI-TOF MS for direct analysis of plant underivatized carbohydrates in subpicoliter single-cell cytoplasm extract. *J Am Soc Mass Spectr* 2008, 19:1841-1848. Publisher Full Text

34. Dreisewerd K, Lemaire R, Pohlentz G, Salzet M, Wisztorski M, Berkenkamp S, Fournier I:Molecular profiling of native and matrix-coated tissue slices from rat brain by infrared and ultraviolet laser desorption/ionization orthogonal time-of-flight mass spectrometry. *Anal Chem* 2007, 79:2463-2471.

35. Sun G, Yang K, Zhao Z, Guan S, Han X, Gross RW: Matrix-assisted laser desorption/ionization time-of-flight mass spectrometric analysis of cellular glycerophospholipids enabled by multiplexed solvent dependent analyte-matrix interactions. *Anal Chem* 2008, 80:7576-7585.

36. Zhang H, Cha S, Yeung ES: Colloidal graphite-assisted laser desorption/ionization (GALDI) MS and MSn of small molecules. 2. Direct profiling and MS imaging of small metabolites from fruits. *Anal Chem* 2007, 79:6575-6584.

37. Hölscher D, Shroff R, Knop K, Gottschaldt M, Crecelius A, Schneider B, Heckel DG, Schubert US, Svatoş A: Matrix-free UV-laser desorption/ionization (LDI) mass spectrometric imaging at the single-cell level: Distribution of secondary metabolites of Arabidopsis thaliana and Hypericum species. *Plant J* 2009, 60:907-918.

38. Li Y, Shrestha B, Vertes A: Atmospheric pressure infrared MALDI imaging mass spectrometry for plant metabolomics. *Anal Chem* 2008, 80:407-420.

39. Shrestha B, Vertes A: In situ metabolic profiling of single cells by laser ablation electrospray ionization mass spectrometry. *Anal Chem* 2009, 81:8265-8271.

40. Dreisewerd K, Draude F, Kruppe S, Rohlfing A, Berkenkamp S, Pohlentz G: Molecular analysis of native tissue and whole oils by infrared laser mass spectrometry. *Anal Chem* 2007, 79:4514-4520.

41. Dreisewerd K, Berkenkamp S, Leisner A, Rohlfing A, Menzel C: Fundamentals of matrix-assisted laser desorption/ionization mass spectrometry with pulsed infrared lasers. *Int J Mass Spectrom* 2003, 226:189-209.

42. Trim PJ, Atkinson SJ, Princivalle AP, Marshall PS, West A, Clench MR: Matrix-assisted laser desorption/ionisation mass spectrometry imaging of lipids in rat brain tissue with integrated unsupervised and supervised multivariant statistical analysis. *Rapid Commun Mass Spectrom* 2008, 22:1503-1509.

43. Henderson A, Fletcher JS, Vickerman JC: A comparison of PCA and MAF for ToF-SIMS image interpretation. *Surf Interface Anal* 2009, 41:666-674.

44. Loboda AV, Ackloo S, Chernushevich IV: A high-performance matrix-assisted laser desorption/ionization orthogonal time-of-flight mass spectrometer with collisional cooling. *Rapid Commun Mass Spectrom* 2003, 17:2508-2516.

45. Krutchinsky AN, Loboda AV, Spicer VL, Dworschak R, Ens W, Standing KG: Orthogonal injection of matrix-assisted laser desorption/ionization ions into a time-of-flight spectrometer through a collisional damping interface. *Rapid Commun Mass Spectrom* 1998, 12:508-518.

46. Soltwisch J, Souady J, Berkenkamp S, Dreisewerd K: Effect of gas pressure and gas type on the fragmentation of peptide and oligosaccharide ions generated in an elevated pressure UV/IR-MALDI ion source coupled to an orthogonal time-of-flight mass spectrometer. *Anal Chem* 2009, 81:2921-2934.

47. Corpas FJ, Barroso JB, Carreras A, Valderrama R, alma JM, Leon AM, Sandalio LM, del Rio LA:Constitutive arginine-dependant nitric oxide synthase activity in different organs of pea seedlings during plant development. *Planta* 2006, 224:246-254.

48. Wendehammer D, Durner J, Klessig DF: Nitric oxide: a new player in plant signaling and defence responses. *Curr Opin Plant Biol* 2004, 7:449-455.

49. Igarashi K, Kashiwagi K: Modulation of cellular function by polyamines. *Int J Biochem Cell Biol* 2010, 42:39-51.

50. Kidd SK, Melillo AA, Lu R-H, Reed DG, Kuno N, Uchida K, Furuya M, Jelesko JG: The A and B loci in tobacco regulate a network of stress response genes, few of which are associated with nicotine biosynthesis. *Plant Molecul Biol* 2006, 60:699-716.

51. Shetty P, Atallah MT, Shetty K: Stimulation of total phenolics, L-DOPA and antioxidant activity through proline-linked pentose phosphate pathway in response to proline and its analogue in germinating fava beans (Vicia faba). *Process Biochem* 2003, 38:1707-1717.

52. Mueller MJ: Archetype signals in plants: the phytoprostanes. *Curr Opin Plant Biol* 2004, 7:441-448.

53. Mueller MJ, Berger S: Reactive electrophilic oxylipins: pattern recognition and signalling. *Phytochemistry* 2009, 70:1511-1521.

54. Shah J: Lipids, Lipases and Lipid-modifying enzymes in plant disease resistance. *Annu Rev Phytopathol* 2005, 43:229-260.

55. Li X-C, Joshi AS, ElSohly HN, Khan SI, Jacob MR, Zhang Z, Khan IA, Ferreira D, Walker D, Walker LA, Broedel SE, Raulli RE, Cihlar RL: Fatty acid synthase inhibitors from plants: isolation, structure elucidation, and SAR studies. *J Nat Prod* 2002, 65:1909-1914.

56. Göbel C, Feussner I, Rosahl S: Lipid peroxidation during the hypersensitive response in potato in the absence of 9-lipoxygenases. *J Biol Chem* 2003, 278:52834-52840.

57. Montillet J-L, Chamnongpol S, Rusterucci C, Dat J, van de Cotte B, Agnel J-P, Battesti C, Inze D, van Breusegem F, Triantaphylidès C: Fatty Acid hydroperoxides and H2O2 in the execution of hypersensitive cell death in tobacco leaves. *Plant Physiol* 2005, 138:1516-1526.

58. Cacas J-L, Vailleau F, Devoine C, Ennar N, Agnel J-P, Tronchet M, Ponchet M, Blein J-P, Roby D, Triantaphylidès C, Montillet J-C: The combined action of 9 lipoxygenase and galactolipase is sufficient to bring about programmed cell death during tobacco hypersensitive response. *Plant Cell Environmen* 2005, 28:1367-1378.

59. Montillet J-L, Cacas J-L, Garnier L, Montané M-H, Douki T, Bessoule J-J, Polkowska-Kowalczyk L, Maciejewska U, Agnel J-P, Vial A, Triantaphylidès C: The upstream oxylipin profile of Arabidopsis thaliana: a tool to scan for oxidative stresses. *Plant J* 2004, 40:439-451.

60. Véronési C, Rickauer M, Fournier J, Pouénat M-L, Esquerré-Tugayé M-T: Lipoxygenase gene expression in the tobacco-phytophthora parasitica nicotianae interaction. *Plant Physiol* 1996, 112:997-1004.

61. Munnik T, Irvine RF, Musgrave A: Phospholipid signalling in plants. *Biochim Biophys Acta* 1998, 1389:222-272.

62. Świątek A, Lenjou M, van Bockstaele D, Inzé D, van Onckelen H: Differential effect of jasmonic acid and abscisic acid on cell cycle progression in tobacco BY-2 cells. *Plant Physiol* 2002, 128:201-211.

63. Seo M, Koshiba T: Complex regulation of ABA biosynthesis in plants. *Trends Plant Sci* 2002, 7:41-48.

64. Steadman JR, Sequeira L: Abscisic acid in tobacco plants. *Plant Physiol* 1970, 45:691-697.

65. Ton J, Flors V, Mauch-Mani B: The multifaceted role of ABA in disease resistance. *Trends Plant Sci* 2009, 14:310-317.

66. Cheng W-H, Endo A, Zhou L, Penney J, Chen H-C, Arroyo A, Leon P, Nambara E, Asami T, Seo M, Koshiba T, Sheen J: A unique short-chain dehydrogenase/reductase in Arabidopsis glucose signaling and abscisic acid biosynthesis and functions. *Plant Cell* 2002, 14:2723-2743.

67. Schmidt K, Plugmacher M, Klages S, Maeser A, Mock A, Stahl DJ: Accumulation of the hormone abscisic acid (ABA) at the infection site of the fungus Cercospora beticola supports the role of ABA as a repressor of plant defence in sugar beet. *Mol Plant Pathol* 2008, 9:661-673.

68. Nambara E, Marion-Poll : Absicic acid biosíntesis and catabolism. *A Annu Rev Plant Biol* 2005, 56:165-185.

69. Cutler AJ, Krochko JE: Formation and breakdown of ABA. *Trends Plant Sci* 1999, 4:472-478.

70. Kenton P, Mur LAJ, Atzorn R, Wasternack C, Draper J: (-)-Jasmonic acid accumulation in tobacco hypersensitive response lesions. *Mol Plant Microbe In* 1999, 12:74-78.

71. Kaiser AE, Gottwald AM, Wiersch CS, Maier WA, Seitz HM: Spermidine metabolism in parasitic protozoa - a comparison to the situation in prokaryotes, viruses, plants and fungi. *Folia Parasit* 2003, 50:3-8.

72. Stork W, Diezel C, Halitschke R, Galis I, Baldwin IT: An ecological analysis of the herbivory-elicited JA burst and its metabolism: plant memory processes and predictions of the moving target model. *PLOS One* 2009, 4: e4697-e4611.

Chapter 8

# Chlorophyll Fluorescence Emission can Screen Cold Tolerance of Cold Acclimated Arabidopsis thaliana Accessions

Anamika Mishra[1], Arnd G Heyer[2] and Kumud B Mishra[1*]

[1]Global Change Research Centre, Academy of Sciences of the Czech Republic, v. v. i, Bělidla 986/4a, 603 00, Brno, Czech Republic

[2]Institute of Biomaterials and Biomolecular Systems, Department of Plant Biotechnology, University of Stuttgart, Stuttgart, Germany

# ABSTRACT


## Background

An easy and non-invasive method for measuring plant cold tolerance is highly valuable to instigate research targeting breeding of cold tolerant crops. Traditional methods are labor intensive, time-consuming and thereby of limited value for large scale screening. Here, we have tested the capacity of chlorophyll *a* fluorescence (ChlF) imaging based methods for the first time on intact whole plants and employed advanced statistical classifiers and feature selection rules for finding combinations of images able to discriminate cold tolerant and cold sensitive plants.

## Results

ChlF emission from intact whole plant rosettes of nine *Arabidopsis thaliana* accessions was measured for (1) non-acclimated (NAC, six week old plants grown at room temperature), (2) cold acclimated (AC, NAC plants acclimated at 4°C for two weeks), and (3) sub-zero temperature (ST) treated (STT, AC plants treated at -4°C for 8 h in dark) states. Cold acclimation broadened the slow phase of ChlF transients in cold sensitive (Co, C24, Can and Cvi) *A. thaliana* accessions. Similar broadening in the slow phase of ChlF transients was observed in cold tolerant (Col, Rsch, and Te) plants following ST treatments. ChlF parameters: *maximum quantum yield of PSII photochemistry* ($F_V/F_M$) and*fluorescence decrease ratio* ($R_{FD}$) well categorized the cold sensitive and tolerant plants when measured in STT state. We trained a range of statistical classifiers with the sequence of captured ChlF images and selected a high performing *quadratic discriminant classifier* (QDC) in combination with*sequential forward floating selection* (SFFS) feature selection methods and found that linear combination of three images showed a reasonable contrast between cold sensitive and tolerant *A. thaliana* accessions for AC as well as for STT states.

## Conclusions

ChlF transients measured for an intact whole plant is important for understanding the impact of cold acclimation on photosynthetic processes. Combinatorial imaging combined with statistical classifiers and feature selection methods worked well for the screening of cold tolerance without exposing plants to sub-zero temperatures. This opens up new possibilities for high-throughput monitoring of whole plants cold tolerance *via* easy and fully non-invasive means.

# BACKGROUND

Cold tolerance is the ability of plants to withstand low temperatures and plays a crucial role in worldwide production of many important agricultural crops. In temperate climate zones, plants have developed strategies to adjust low temperature tolerance by employing a highly complex process of physiological re-arrangements, termed cold acclimation that is triggered by low but non-freezing temperatures. Numerous studies revealed that cold acclimation not only programs massive changes in transcriptome and metabolome [1-7] but also induces structural and compositional modifications of compatible solutes in various sub-cellular compartments [8]. Following acclimation, plants are more efficient in dealing with the impact of sustained cold or sudden temperature drops by activating a plethora of adjustments including accumulation of cryoprotective metabolites and proteins for their survival [9]. A correlation of cold acclimation capacities with habitat winter temperatures points to high metabolic or ecological costs of these adjustments [5,10,11]. Despite intensive research the molecular mechanisms of cold tolerance are still not fully understood and remain an area for intensive research, as understanding of mechanisms responsible for cold acclimation would allow breeding of cold tolerant crops.

The primary requirement for research associated with engineering of low temperature tolerant crops is to develop efficient and cost effective methods for measuring cold tolerance. Measurement of survival or re-growth scores following triggers of cold acclimation and de-acclimation treatments are quite lengthy and may not be very accurate [12,13].

Most quantitative methods for assessing cold tolerance employ analysis of damage to plasma membranes by electrolyte leakage (EL) screens or the thylakoids *via* plastocyanin release after freeze-thaw cycles [14]. Generally, the temperature affording half of maximal damage ($LT_{50}$), i.e. the temperature at which 50% of electrolytes or plastocyanin are released from their respective compartments, is being evaluated and used as a proxy for cold tolerance. However, since the measurement of purified plastocyanin is comparatively difficult [15], the most popular method to study plant cold tolerance is the evaluation of EL of detached leaves by using conductivity measurements [5,16]. According to this method, detached leaves are placed in reaction tubes, which are progressively cooled down to certain temperatures. Low-temperature treated samples are then thawed at 4°C and electrical conductivity of a bathing solution is measured to calculate $LT_{50}$ values that represent half-maximum damage to the plasma membrane [16]. EL is a widely accepted method for quantification of plant cold tolerance, although its results sometimes deviate from those obtained with alternative methods [17,18].

The response of thylakoids membranes to sustained low-temperature treatments can be measured non-invasively by the analysis of ChlF emission. ChlF is re-emitted by chlorophyll *a* molecules following light absorption, and it is modulated by photochemical and non - photochemical events in the photosynthetic pigment–protein complexes PSII and PSI of the thylakoid membranes [19,20]. The relative contribution of PSII and PSI in ChlF emission is also reported to change during cooling treatments [21]; thereby, this method can provide important insights into molecular processes of cold acclimation. The ChlF parameter $F_V/F_M$ of detached leaves following freeze - thaw cycles has successfully been applied to quantify cold tolerance in *Arabidopsis thaliana* as well as in other plant species [reviewed in [8,22-24]. The analysis of polyphasic fluorescence rise of from initial low fluorescence ($F_O$) to peak $F_P$ by JIP test [25] has also been reported to be useful for the selection of cold tolerance in wheat genotypes [26,27]. Earlier, we measured the ChlF transients of detached leaves of differentially cold tolerant *A. thaliana* accessions during progressive cooling starting from room temperature to -15°C and found high correlation between $LT_{50}$ measured by electrolyte leakage with ChlF parameters such as $F_O$(-15°C)/$F_O$(4°C) and $[F_S/F_O]_{-15°C}$[10]. We demonstrated that application of advanced statistics-based combinatorial imaging methods to the sequence of time resolved ChlF images could be used to categorize cold tolerance levels by training of classifiers using fluorescence emission of detached leaves that were slowly cooled at mild sub-zero-temperatures of around -4°C [10]. Vaclavik et al. [7] reported that cold acclimation induced accumulation of Gluconapin and Flavon-3-ol glycosides, respectively, in cold tolerant vs. cold sensitive *A. thaliana*accessions, and thus metabolomic patterns, could be used for the screening of cold tolerance already in the cold acclimated state i.e. without exposing plants to freezing temperatures. Although metabolomic based methods can thus have better discriminating capacity of cold tolerant and cold sensitive accessions as compared to EL methods, they are not readily applicable in high-throughput screenings. Therefore, in the search for easy methods of sensing cold tolerance, we have tested the potential of ChlF transients measured on whole plant rosettes of the model species *Arabidopsis thaliana*. We involved NAC, AC, and STT plants of nine *A. thaliana* accessions that span the north–south range of the species [5]. A short light/dark protocol was applied for recording of ChlF transients of dark-adapted plants for about 202 s using a pulse amplitude modulation (PAM) based fluorometer[28]. We found that ChlF transients averaged over whole intact plant rosettes offered valuable information for assessing impacts of cold acclimation and cold induced alterations of the photosynthetic machinery. In addition to classical analysis of ChlF transients, we employed statistical classifiers and feature selection methods on captured ChlF images to search for highly contrasting features of cold tolerant vs. cold sensitive accessions. We have trained several classifiers and chose the best performing classifier, i.e. a QDC, for further use along with a high performing SFFS feature selection methods to identify features correlated with cold tolerance. The sets of images obtained for the STT state were then tested on AC state data sets, and we found that this method, i.e. combinatorial imaging, can be applied for assessing plant cold tolerance already in the cold acclimated state, i.e. without any sub-zero temperature treatments and by fully non-invasive means.

# RESULTS

## ChlF Transients from Whole Plants are Highly Informative for Categorization of Cold Tolerance in *A. thaliana* Accessions

We measured the ChlF transients of whole plant rosettes in nine differentially cold tolerant*Arabidopsis thaliana* accessions in the NAC, AC, and STT state [Figure 1]. From the shape of ChlF transients it can be established that for the cold sensitive accessions (Co, C24, Can, and Cvi), the fluorescence intensity following peak $F_P$ quenched slowly in cold acclimated (AC) state, whereas the value of $F_P$ itself decreased significantly when these plants were treated with sub-zero temperature (ST, -4°C for 8 h in dark). In cold tolerant accessions (Te, Rsch, and Col-0), however, the shape of the ChlF transients remained similar for NAC and AC states; while after STT, quenching of fluorescence intensity following peak $F_P$ is significantly slowed down. Qualitative comparison of the ChlF transients of intermediate accessions (Ler and Nd) revealed a compound picture, and based on qualitative comparisons we tend to assign Ler as cold sensitive, while Nd can be classified cold tolerant.

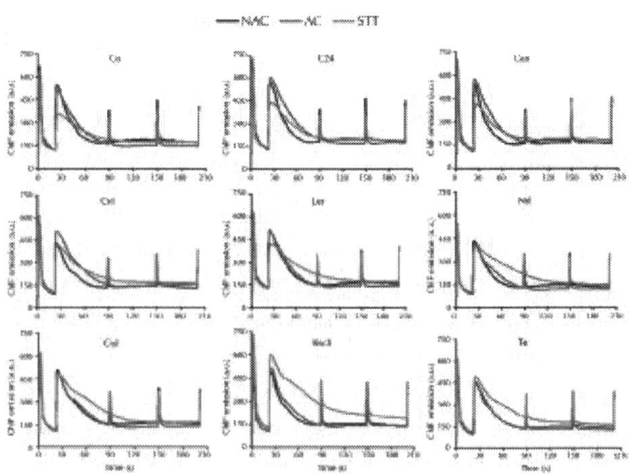


**Figure 1:** The chlorophyll fluorescence (ChlF) transients of nine differentially cold tolerant *Arabidopsis thaliana* accessions measured by short experimen-

tal light/dark protocol. The cold sensitive accessions are Co, C24, Can and Cvi (d); intermediate cold tolerant accession are Ler and Nd; and cold tolerant accessions are Col, Rsch and Te. The ChlF transient was integrated across whole plant rosettes and mean of three independent plants for each accession was presented.

## ChlF Parameters $F_V/F_M$ and $R_{FD}$ of Whole Plant Rosettes can Measure Cold Tolerance in Subzero-Temperature (ST) Treated Plants

Fluorescence parameters $F_V/F_M$ and $R_{FD}$ were evaluated from measured ChlF emission and presented as an average of three independent experiments, where values of each parameter were estimated by integrating across whole rosette leaves. We did not find correlations between the classical ChlF parameters and cold tolerance among the investigated *A. thaliana* accessions either for NAC or for the AC state. However, the ChlF parameters $F_V/F_M$ and $R_{FD}$, differed for cold sensitive, intermediate and cold tolerant accessions following STT (Figure 2). In STT state, $F_V/F_M$ values for cold tolerant (Te, Col and Rsch), intermediate (Nd and Ler) and sensitive accessions (Co, C24, and Can) were around $0.85 \pm 0.01$, $0.83 \pm 0.01$ and $0.81 \pm 0.01$, respectively, while the averaged $R_{FD}$ values were calculated as $1.98 \pm 0.19$, $1.48 \pm 0.07$ and $1.40 \pm 0.19$, respectively. Non-paired t-test revealed $F_V/F_M$ to differ significantly ($p < 0.01$) for cold tolerant and cold sensitive accessions, while there was no significant difference between cold sensitive and intermediate or intermediate and tolerant accessions. In contrast, plant vitality index $R_{FD}$ showed significant differences between the cold sensitive and tolerant as well as between intermediate and tolerant accessions ($p < 0.05$), but no significant difference was found between sensitive and intermediate tolerant accessions. For the accessions Cvi the values of ChlF parameters $F_V/F_M$ and $R_{FD}$ are almost similar to that of intermediate accessions Ler and Nd, thereby its tolerance level of thylakoid may be categorized intermediate, while the plasma membrane behaved cold sensitive in EL measurements.

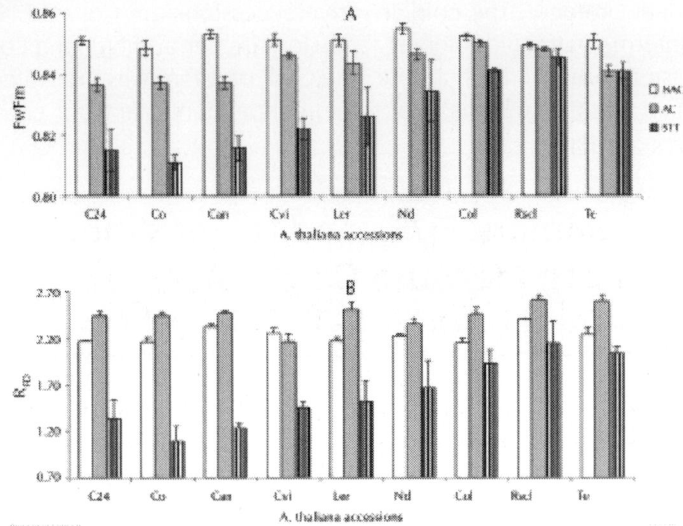


**Figure 2:** Chlorophyll fluorescence (ChlF) parameters*: maximum quantum yield of PSII photosystems* (FV/FM) [A] and *fluorescence decrease ratio* (RFD) [B] of differentially cold tolerant accessions of*Arabidopsis thaliana* for non-acclimated (NAC), cold acclimated (AC), and sub-zero temperature treated (STT) states. The presented numeric values are mean of three independent plants with standard errors, and are integrated across whole plant rosettes.

## Cold Acclimation Induced Gain in Photosynthetic Performance of *A. thaliana* Accessions

The ChlF parameter *effective quantum efficiency of PS II photochemistry* ($\Phi_{PSII}$) was significantly higher ($p < 0.01$) in AC vs. NAC plants except for Cvi (Figure 3A). Interestingly, mild sub-zero temperature treatments (-4°C) for eight hours in AC plants led to a significant decline in the value of $\Phi_{PSII}$ in all accessions (Figure 3A) without $\Phi_{PSII}$ being correlated with the cold tolerance. Two weeks of cold acclimation caused a significant ($p < 0.01$) increase of *photochemical quenching* (qP) in all accessions, while STT led to a decline (Figure 3B). As for $\Phi_{PSII}$, no correlation of parameter qP with cold tolerance was observed.

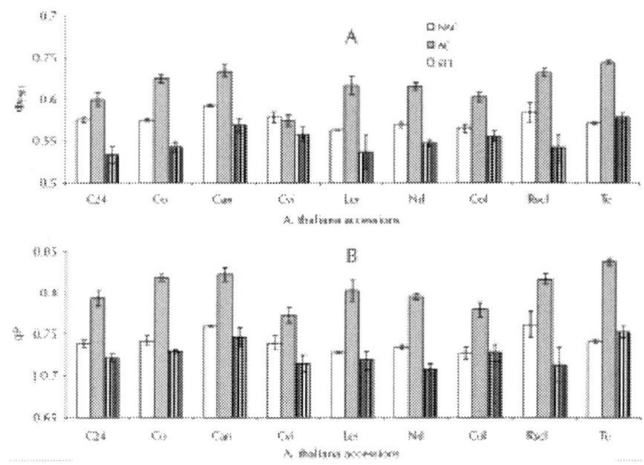


**Figure 3:** The *effective quantum efficiency PS II* (**Φ**PSII) [A] and*photochemical quenching* (qP) [B] of cold sensitive (Co, C24, Cvi, Can), intermediate cold tolerant (Nd and Ler) and cold tolerant (Te, Rsch and Col) *Arabidopsis thaliana* accessions for non-acclimated (NAC), cold acclimated (ACC), and sub-zero temperature treated (STT) states. The presented numeric values are integrated across whole plant rosettes and mean of three independent plants with standard errors.

## Combinatorial Imaging

We applied statistical techniques of classifier and feature selection methods in order to identify features of cold tolerance from captured sequences of time-resolved ChlF images. This method is very powerful in identifying image sets from large sequences of time-resolved ChlF recordings that yield the highest contrast between groups to be compared. Earlier we demonstrated the usefulness of this approach for discriminating three species of family *Lamiaceae* at very early stages of growth [29] and for investigating features of cold tolerance at non-lethal temperatures [10]. The method depends on training and performance testing of randomly selected pixels of the most contrasting ChlF image sets, i.e. recordings for the highly cold tolerant accession Te and highly sensitive Co in the STT state [for technical details, see [10,29,30]. The performance, error rate and computational time to run the algorithms of five tested classifiers, *linear discriminant classifier* (LDC), *quadratic discriminant classifier* (QDC), *k-nearest neighbors classifier* (*k*-NNC), *nearest neighbor classifier* (NNC) and*nearest mean classifier* (NMC), are presented in Table 1. We found that algorithms of QDC are the best performing classifier (81% correct assignment of cold tolerance group among test images) in comparatively short time (<6 s, Table 1). Therefore, QDC was chosen and applied with the SFFS feature selection method to find the most contrasting image sets for highly cold tolerant (Te) vs. highly cold sensitive (Co) accessions. Figure 4 shows the performance curve for the most efficient feature selection method SFFS in combination with simulated classifier QDC, where x-axis represents number of images. Thus, in this experiment, the SFFS algorithm reduced the full data set of 218 images to three images identified as $I_{21,}$ $I_{104}$ and $I_{107}$ without compromising the classification performance (~80%, Figure 4). We obtained the linear combination: $C = (0.3219)*I_{21} + (-0.5018)*I_{104} + (-0.2315)*I_{107}$, to perform best in discriminating cold tolerant and sensitive accessions. The coefficients of the linear combination were calculated according to Matouš et al. [30]. Figure 5 shows linear combinations of images for all nine accessions in the STT as well as AC state. The presented images are plotted on a false color scale where color represents the virtual fluorescence intensity of the pixel with reference to its corresponding value in the training data set. Thus, we can visualize a clear difference between cold sensitive, intermediate, and cold tolerant accessions in the STT state. When the combinatorial imaging was applied to AC plants it apparently failed to discriminate cold sensitive and intermediate accessions showing very similar patterns for leaf rosettes of the two groups of plants. However, the tolerant group of accessions could be clearly discriminated already in the AC state.

**Table 1:** The table presents performance, error rate and computational time of the tested classifiers

| Classifiers | Performance | Error rate | Computational time (s) |
|---|---|---|---|
| LDC | 0.81 | 0.09 | 6.14 |
| QDC | 0.81 | 0.09 | 5.86 |
| *k*-NNC (*k* = 3) | 0.82 | 0.08 | 5530 |
| NNC (k = 1) | 0.79 | 0.10 | 5510 |
| NMC | 0.47 | 0.26 | 8.23 |

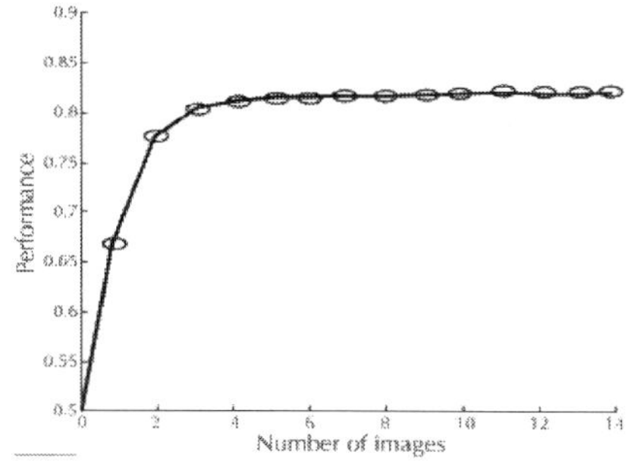


**Figure 4:** The performance curve of *sequential forward floating selection* (SFFS) with images in the x-axis.

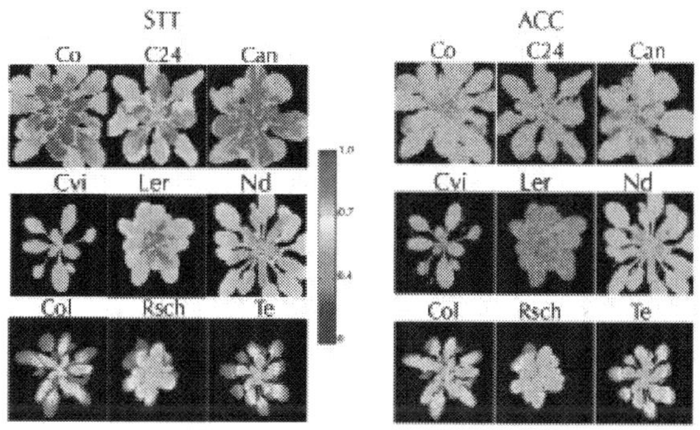


**Figure 5:** Illustrates the combination of three most contrasting images for all nine accessions of *Arabidopsis thaliana*. After subzero temperature treatment the all nine accession has been visibly divided in three categories namely, most sensitive (Co, C24 and Can), intermediates (Ler and Nd) and the most tolerant accessions (Te, Rsch and Col).

# DISCUSSION

## Towards Easy Methods for Assessing Whole Plant Cold Tolerance in Natural Accessions of *A. thaliana*

In an attempt to establish an easy and non-invasive method for screening plant cold tolerance, we compared different concepts for measuring cold tolerance based on chlorophyll fluorescence that have been approved for detached leaves of the model species *Arabidopsis thaliana*. As far as we know, for the first time we attempted to capture the ChlF transients of intact whole plant rosettes and found interesting modulation in it for the group of cold sensitive (Co, C24, Can, Cvi), intermediate tolerant (Ler, Nd) and cold tolerant (Te, Rsch, Col-0) *A. thaliana* accessions for NAC, AC and STT state (Figure 1). The ChlF transient measured by quenching analysis of modulated fluorescence is very useful methods that can monitor slight changes in the photochemical and non-photochemical process[31]. The slow quenching of ChlF transients following peak $F_P$, i.e. broadening of ChlF transients, for cold sensitive group of accessions under cold acclimated state and for tolerant group of accessions under STT state (Figure 1) is complex and can be affected by series of events such as light induced intrathylakoid $H^+$ concentration, non-photochemical quenching, inhibition of $CO_2$ assimilation processes, ATP synthesis among others [reviewed in [32]. However, it reveals that sensitive and tolerant group of accessions possess two different strategies to utilize the absorbed irradiance following cold acclimation as well as when placed at mild subzero temperature (-4°C) for 8 h in darkness. Probably cold tolerant plants efficiently utilize acclimation temperature and keeping homeostasis of photosynthetic apparatus and yielded almost similar ChlF transients in NAC as well as in AC state that varied only when treated with sub-zero temperature. The reduction of $F_P$ for the group of cold sensitive accessions in STT state indicates perturbation in PSII functional properties. Significant inhibition of $F_V/F_M$ for sensitive vs. tolerant accessions in STT states further support slight inactivation of PSII function. Ehlert and Hincha [23] reported that the easily recordable parameter $F_V/F_M$ discriminates cold tolerant and sensitive detached leaves after a freeze-thaw cycle. Using a set of nine differentially cold tolerant natural accessions of *Arabidopsis*[5], we could verify also for whole plants that $F_V/F_M$ differs for tolerant and sensitive accessions after a sub-zero temperature treatment, but plants with intermediate tolerance differed neither from sensitive nor from tolerant ones. It has been reported that the fluorescence decrease ratio, $R_{FD}$, is more sensitive to a variety of stress factors than $F_V/F_M$, which mainly responds to extreme conditions [33,34]. Indeed, $R_{FD}$, which can be calculated from fluorescence peak $F_P$ and $F_S$ measured under prevailing actinic light, was not only able to discriminate sensitive and tolerant but also tolerant and intermediate accessions. Interestingly, both $F_V/F_M$ and $R_{FD}$ data revealed that plastids of the accession Cvi behaved as intermediately tolerant, while the electrolyte leakage method classified this accession as sensitive based on damage to the plasma membrane during freeze-thaw treatments [5]. We have earlier reported that cold tolerance of plastids and the plasma membrane appear not strictly correlated in the accession Col-0 [8], and the differential tolerance of plastids and plasma membrane in Cvi is further evidence for independent principles underlying cold tolerance of the different types of membranes.

Considering plants not treated with sub-zero temperatures, which would be desirable for a cold tolerance screening program, neither $F_V/F_M$ nor $R_{FD}$ were correlated with tolerance of the nine*Arabidopsis* accessions used in this study. Since both parameters are related to intactness of PSII, it is not unexpected that they are not affected by above-zero temperatures in a cold tolerant plant like*Arabidopsis*. However, it was noticed that parameters $\Phi_{PSII}$, and qP that reflect effects of actinic light on chlorophyll fluorescence responded differentially to low above-zero and sub-zero temperature treatments. It has been reported that shifting plants from normal to low temperatures causes repression of genes related to photosynthesis [35], and among them are genes directly involved in light harvesting like e.g. proteins Lhca2*1 and Lhcb4*2 [36]. Down-regulation of photosynthesis genes is accompanied by a sudden suppression of photosynthesis at low temperatures [37]. However this might be a transient effect, since leaves that developed at low temperatures are reported to regain full photosynthetic activity [38]. Although a reduction in antenna size would not necessarily impact parameters $\Phi_{PSII}$ and qP, it is not self-evident why low temperatures should increase effective quantum yield. Savitch & co-workers [38] reported that low temperatures could increase photosynthetic capacity in species like e.g. wheat that maintain a high need for assimilates because of active growth at low temperatures. In the cold tolerant tomato species *Lycopersicon peruvianum*, a rise in qP during cold exposure was hypothesized to result from increased capacity of the Calvin-Benson cycle [39]. In this context it should be noted that temperature effects on light absorption and photosynthesis are often not easily separable from light effects, because in most studies light intensity is reduced at low temperature to avoid damage to the photosystems in the cold. At lower light intensity, qN is reduced and qP rises, and in turn also $\Phi_{PSII}$ is increased as long as no damage of the photosystems occurs. This effect may superimpose the temperature effect on light harvesting, and although the cold sensitive accession Cvi could be discriminated from all intermediate and tolerant accessions based on the lack of increase of $\Phi_{PSII}$ during acclimation, this parameter does not seem to be sufficient for screening cold tolerance in acclimated *Arabidopsis* plants.

## Combinatorial Imaging can Discriminate Cold Acclimation Induced Cold Tolerant Plants

We have earlier reported that combinatorial imaging of ChlF transients combined with classifier and feature selection methods was able to discriminate detached leaves from cold sensitive, tolerant and intermediate accessions of Arabidopsis [10]. The main outcome of the current study is that this method can also be applied to whole plants, i.e. leaves to be compared need not be present on the same images, and thus this method is suitable for large scale screening of plant cold tolerance.

Combinatorial imaging combines sets of highly performing images from sequences of time-resolved ChlF images that provide strong discrimination capacity [10,29,30]. High performing images extract information from several thousands of pixels from hundreds of measured images (218 images per data for this experiment, and each image with resolution 512*512 pixels). This method does not trace underlying physiological phenomena rather its algorithms select image sets having optimal contrast between sensitive and tolerant features that ultimately allow discrimination on the basis of their cold tolerance levels [30]. Therefore, parameters identified for the classifiers are very likely species and treatment specific, and training of the classifiers needs to be performed for each application of the method. While this might appear disadvantageous, it also offers great flexibility, because successful discrimination is not dependent on any specific physiological process but can exploit every feature that allows discrimination [10,29,30,40,41]. In the current study, older leaves performed better than developing ones, probably because the higher fluorescence intensity provided a better signal-to-noise ratio. However, this does not implicate that the method is not suited to the screening of plant seedlings, because it analyzes relative fluorescence intensities distributed among the image pixels and for each sample type it is different [29]. Therefore, combinatorial imaging out-performs existing methods that depend on the application of freeze-thaw treatments.

The combinatorial imaging method could discriminate sensitive and tolerant accessions in the cold acclimated state but it failed to make any difference between cold sensitive and intermediate accessions. Because training was done for STT plants, this failure could result from differential responses of the photosynthetic apparatus to low above-zero and sub-zero temperatures as outlined above, or it could relate to insufficient resolution, which might be overcome by a comparison of larger sets of images. Since this would clearly result in higher computational load, a case specific decision for higher resolution vs. shorter analysis time might be necessary.

Even if only a two-class discrimination can be achieved with the combinatorial imaging method, it would be very useful for large scale screening of cold tolerance e.g. in recombinant inbred line populations or other plant sets consisting of large numbers of individuals representing different genotypes. The method is thus well suited for quantitative trait loci (QTL) mapping or mutant screenings to investigate genetic determinants of cold tolerance that may be used for plant breeding.

## CONCLUSIONS

We have demonstrated that chlorophyll fluorescence emission from whole plant rosettes of*Arabidopsis thaliana* integrates information that can be used to discriminate cold sensitive and tolerant plants in the cold acclimated state when analyzed by the advanced statistics based combinatorial imaging approach [10,29,30]. This reveals the power of combinatorial imaging for identifying features of cold tolerant and sensitive accessions at cold acclimation state (Figure 5) where well known physiological parameters of ChlF emission (Figures 2 and 3) failed to provide any clue of discrimination. Moreover, capturing ChlF transients of whole plant rosettes following mild sub-zero temperature treatments (STT, -4°C for ~8 h in dark) is also very useful, because classical ChlF transients and the extracted parameters such as $F_V/F_M$ and $R_{FD}$ can categorize cold sensitive, intermediate and tolerant accessions following STT. In addition, classical ChlF analysis, in contrast to combinatorial imaging methods, can yield physiologically relevant information that could be directly exploited for breeding efforts. The screening of whole plants cold tolerance *via* fully non-invasive means following cold acclimation, i.e. without any sub-zero temperature treatments, could be highly useful in high-throughput screening of cold tolerance where it is superior to data measured from single leaf or leaf discs by alternative methods such as metabolomics [7] or electrolyte leakage [5].

# METHODS

## Plant Material and Growth Conditions

Six plantlets of each of the nine accessions of *Arabidopsis thaliana*[5] were grown inside a cooling chamber in pots of 0.06 m for six weeks at day/night temperatures 20°C/18°C, light 90 μmol (photons) $m^{-2}$ $s^{-1}$ and a relative humidity of 70%. The accessions used in this study were: Cvi (Cap Verde Islands), Can (Canary Islands), C24 (genetically related to Co, Portugal), Co (Coimbra, Portugal), Col-0 (Columbia-0, genetically related to Gü, Germany), Nd (Niederzenz, Germany), Ler (Landsberg *erecta,* Poland), Rsch (Rschew, Russia), and Te (Tenela, Finland). The experiments were executed in three independent replica with three independent sets of plants used to measure ChlF in the non acclimated (NAC, six weeks grown plants), cold acclimated (AC, NAC plants acclimated at 4°C for another 2 weeks), and sub-zero temperature treated (STT, AC plants treated with mild sub-zero temperature of -4°C for 8 hours in dark) state.

## Chlorophyll Fluorescence Measurement

Individual plants were used for ChlF measurements at room temperature using Handy FluorCam [http://www.psi.cz *webcite*; [23]. A short protocol of ~202 seconds modified according to Mishra et al. [10] was used for capturing time-resolved ChlF from plant rosettes. This protocol starts with the measurement of basal fluorescence ($F_O$) and maximum fluorescence ($F_M$) using measuring and saturating flashes. After a short dark period of ~20 seconds, actinic light of 40 μmol (photons) $m^{-2}$ $s^{-1}$ was applied to measure the fluorescence transients. Two strong flashes of saturating light were overlaid with actinic light to investigate activation of non-photochemical quenching followed by a third saturating flash 18 s after switching off the actinic light to see the relaxation of non-photochemical quenching mechanisms. Three replicate recordings were taken for NAC, AC and STT state, respectively. The images of the measured ChlF transients were averaged across the whole rosettes for quantitative evaluation of the fluorescence parameters or plotting of ChlF transients.

## Combinatorial Imaging Using Statistical Classifier and Feature Selection Method

The method of combinatorial imaging was applied to identify the discriminant between the accessions without using the whole data set consisting of 218 images in a time series. Each time series contains some repetitive images as well as images with low contrast, which were sorted out to reduce the size of the data sets. Application of statistical techniques of classifiers was chosen to avoid biasness. We randomly classified the data for Te and Co as training and testing sets. Using the training set, features discriminating the tolerant and sensitive accession were identified, and these features were next applied using the testing set for discriminating cold tolerant and sensitive accessions. Using this method we calculated the performance of several classifiers: LDC, QDC, *k*-NNC, NNC and NMC, and chose the best performing classifier for further analysis. The performance of each of the investigated classifiers was quantified by a number between 0 1: value'0' means random classification (1/2 of classifications into 2 equally represented classes correct, and another 1/2 is incorrect) and value '1' meaning that the classifier was 100% successful [for details see [24]. In this experiments performance of QDC was 81% and computational time was less than 6 seconds. Therefore, QDC classifier was applied with sequential forward floating selection (SFFS) to reduce the number of images for effective classification [42]. The reduction is based on finding an image sub-set containing the most useful information for the visualization of highly contrasting features of cold tolerant vs. sensitive accessions and arranging the images in a form of descending order of their performance efficiency. The combinatorial images are developed by the combination of the three high performing images (x,y,z) multiplied with their coefficient (a,b,c) that are obtained by *linear discriminant analysis*(LDA): Combinatorial imaging (C) = $(\pm a)*I_x + (\pm b)*I_y + (\pm c)*I_z$.

### Tool for the Data Analysis

The Matlab software package, version 6.5, with Pattern Reorganization toolbox (PRTools) was used for statistical analysis.

## ACKNOWLEDGEMENTS

Participation of AM and KBM was supported by the EfCOP - IPo project ENVIMET (CZ.1.07/2.3.00/20.0246) and by the EU (FP7) EPPN project (284443). We thank Ladislav Nedbal for his inspiration.

## REFERENCES

1. Jaglo-Otteson KR, Gilmour SJ, Zarka DG, Schabenberger O, Thomashow MF: Arabidopsis CBF1 overexpression induces COR genes and enhances freezing tolerance. *Science* 1998, 280:104-106.
2. Cook D, Fowler S, Fiehn O, Thomashow MF: A prominent role for the CBF cold response pathway in configuring the low-temperature metabolomes of Arabidopsis. *Proc Natl Acad Sci U S A* 2004, 101(45):15243-15248.
3. Kaplan F, Kopka J, Haskell DW, Zhao W, Schiller KC, Gatzke N, Sung DY, Guy CL: Exploring the temperature-stress metabolome of Arabidopsis. *Plant Physiol* 2004, 136:4159-4168.
4. Guy C, Kaplan F, Kopka J, Selbig J, Hincha DK: Metabolomics of temperature stress. *Physiol Plant* 2008, 132:220-235.
5. Hannah MA, Wiese D, Freund S, Fiehn O, Heyer AG, Hincha DK: Natural genetic variation of freezing tolerance in Arabidopsis. *Plant Physiol* 2006, 142:98-112.
6. Zuther E, Schulz E, Childs LH, Hinch DK: Clinal variation in the non-acclimated and cold-acclimated freezing tolerance of *Arabidopsis thaliana* accessions. *Plant Cell Environ* 2012, 35:1860-1878.

7. Vaclavik L, Mishra A, Mishra KB, Hajslova J: Mass spectrometry-based metabolomic fingerprinting for screening cold tolerance in *Arabidopsis thaliana* accessions. *Anal Bioanal Chem* 2013, 405(8):2671-2683.
8. Knaupp M, Mishra KB, Nedbal L, Heyer AG: Evidence for a role of raffinose in stabilizing photosystem II during freeze–thaw cycles. *Planta* 2011, 234(3):477-486.
9. Moellering ER, Bagyalakshmi M, Benning C: Freezing tolerance in plants requires lipid remodeling at the outer chloroplast membrane. *Science* 2010, 330:226-228.
10. Mishra A, Mishra KB, Höermiller II, Heyer AG, Nedbal L: Chlorophyll fluorescence emission as a reporter on cold tolerance in *Arabidopsis thaliana* accessions. *Plant Signal Behav* 2011, 6(2):301-310.
11. Degenkolbe T, Giavalisco P, Zuther E, Seiwert B, Hinchey DK, Willmitzer L: Differential remodeling of the lipidome during cold acclimation in natural accessions of*Arabidopsis thaliana*. *Plant J* 2012, 72:972-982.
12. Rapacz M, Gasior D, Zwierzykowski Z, Lesniewska-Bocianowska A, Humphreys MW, Gay AP:Changes in cold tolerance and the mechanisms of acclimation of photosystem II to cold hardening generated by anther culture of Festuca pratensis x Lolium multiflorum cultivars. *New Phytol* 2004, 162(1):105-114.
13. Novillo F, Alonso JM, Ecker JR, Salinas J: CBF2/ DREB1C is a negative regulator of CBF1/DREB1B and BF3/DREB1A expression and plays a central role in stress tolerance in Arabidopsis. *Proc Natl Acad Sci U S A* 2004, 101:3985-3990.
14. Hincha DK: Cryoprotectin: a plant lipid-transfer protein homologue that stabilizes membranes during freezing. *Philos T R Soc B* 2002, 357:909-915.
15. Hincha DK, Hofner R, Schwab KB, Heber U, Schmi JM: Membrane rupture is the common cause of damage to chloroplast membranes in leaves injured by freezing or excessive wilting. *Plant Physiol* 1987, 83:251-253.
16. Steponkus PL, Lynch DV, Uemura M: The influence of cold-acclimation on the lipid-composition and cryobehavior of the plasma-membrane of isolated rye protoplasts. *Philos T R Soc B* 1990, 326:571-583.

17. Azzarello E, Mugnai S, Pandolfi C, Masi E, Marone E, Mancuso S: Comparing image (fractal analysis) and electrochemical (impedance spectroscopy and electrolyte leakage) techniques for the assessment of the freezing tolerance in olive. *Trees* 2009, 23:159-167.
18. Yamori W, Noguchi K, Terashima I: Temperature acclimation of photosynthesis in spinach leaves: analyses of photosynthetic components and temperature dependencies of photosynthetic partial reactions. *Plant Cell Environ* 2005, 28:536-547.
19. Govindjee: Sixty-three years since Kautsky: chlorophyll a fluorescence. *Aust J Plant Physiol* 1995, 22:131-160.
20. Baker N: Chlorophyll fluorescence: a probe of photosynthesis *in vivo*. *Annu Rev Plant Biol* 2008, 59:89-113.
21. Agati G, Cerovic ZG, Moya I: The effect of decreasing temperature up to chilling values on the *in vivo* F685/F735 chlorophyll fluorescence ratio in *Phaseolus vulgaris* and*Pisum sativum*: The role of the photosystem I contribution to the 735 nm fluorescence band. *Photochem Photobiol* 2000, 72:75-84.
22. Baker NR, Rosenqvist E: Applications of chlorophyll fluorescence can improve crop production strategies: an examination of future possibilities. *J Exp Bot* 2004, 55(403):1607-1621.
23. Ehlert B, Hincha DK: Chlorophyll fluorescence imaging accurately quantifies freezing damage and cold acclimation responses in Arabidopsis leaves. *Plant Methods* 2008, 4:12.
24. Thalhammer A, Hincha DK, Zuther E: Measuring freezing tolerance: electrolyte leakage and chlorophyll fluorescence assays. *Plant Cold Acclimation: Methods Mol Biol* 2014, 1166:15-24.
25. Strasser RJ, Srivastava A, Govindjee: Polyphasic chlorophyll-Alpha fluorescence transcient in plants and cyanobacteria. *Photochem Photobiol* 1995, 61(1):32-42.
26. Rapacz M, Gasior D, Koscielniak J, Kosmala A, Zwierzykowski Z, Humphreys MW: The role of the photosynthetic apparatus in cold acclimation of Lolium multiflorum. Characteristics of novel genotypes low-sensitive to PSII over-reduction. *Acta Physiol Plant* 2007, 29(4):309-316.

27. Rapacz M, Wozniczka A: A selection tool for freezing tolerance in common wheat using the fast chlorophyll a fluorescence transient. *Plant Breed* 2009, 128(3):227-234.

28. Nedbal L, Soukupová J, Kaftan D, Whitmarsh J, Trtílek M: Kinetic imaging of chlorophyll fluorescence using modulated light. *Photosynth Res* 2000, 66:3.

29. Mishra A, Matouš K, Mishra KB, Nedbal L: Towards discrimination of plant species by machine vision: advanced statistical analysis of chlorophyll fluorescence transients. *J Fluoresc* 2009, 19:905-913.

30. Matouš K, Benediktyova Z, Berger S, Roitsch T, Nedbal L: Case study of combinatorial imaging: what protocol and what chlorophyll fluorescence image to use when visualizing infection of Arabidopsis thaliana by pseudomonas syringae? *Photosynth Res* 2006, 90:243-253.

31. Maxwell K, Johnson GN: Chlorophyll fluorescence: a practical guide. *J Exp Bot* 2000, 51:659-668.

32. Papageorgiou GC, Govindjee: Photosystem II fluorescence: slow changes – scaling from the past. *J Photoch Photobio B* 2011, 104(1–2):258-270.

33. Lichtenthaler HK, Langsdorf G, Lenk S, Buschmann C: Chlorophyll fluorescence imaging of photosynthetic activity with flash-lamp fluorescence imaging system. *Photosynthetica* 2005, 43:355-369.

34. Horgan DB, Zabkiewicz JA: Fluorescence decline ratio: comparison with quantum yield ration for plant physiological status and herbicide treatment responses. *New Zealand Plant Protection* 2008, 61:169-173.

35. Hannah MA, Heyer AG, Hincha DK: A global survey of gene regulation during cold acclimation in *Arabidopsis thaliana.PLoS Genet* 2005, 1(2):e26.

36. Fowler F, Thomashow MF: Arabidopsis transcriptome profiling indicates that multiple regulatory pathways are activated during cold acclimation in addition to the CBF cold response pathway. *Plant Cell* 2002, 14(8):1.

37. Strand A, Hurry V, Gustafsson P, Gardeström P: Development of *Arabidopsis thaliana*leaves at low temperature releases

the suppression of photosynthesis and photosynthetic gene expression despite the accumulation of soluble carbohydrates. *Plant J* 1997, 12:605-614.

38. Savitch LV, Leonardos ED, Krol M, Jansson S, Grodzinski B, Huner NPA, Oquist G: Two different strategies for light utilization in photosynthesis in relation to growth and cold acclimation. *Plant Cell Environ* 2002, 25:761-771.
39. Linger P, Brüggemann W: Correlations between chlorophyll fluorescence quenching parameters and photosynthesis in a segregating Lycopersicon esculentum × L. peruvianum population as measured under constant conditions. *Photosynth Res* 1999, 61:145-156.
40. Codrea MC, Hakala-Yatkin M, Kårlund-Marttila A, Nedbal L, Aittokallio T, Nevalainen OS, Tyystjärvi E: Mahalanobis distance screening of Arabidopsis mutants with chlorophyll fluorescence. *Photosynth Res* 2010, 105(3):273-83.
41. Mattila H, Valli P, Pahikkala T, Teuhola J, Nevalainen OS, Tyystjärvi E: Comparison of chlorophyll fluorescence curves and texture analysis for automatic plant identification. *Precis Agric* 2013, 14:621-636.
42. Jain AK, Duin RPW, Mao JC: Statistical pattern recognition: a review. *IEEE Trans Pattern Anal Mach Intell* 2000, 22(1):4-37.

Chapter 9

# LEAF-E: A Tool to Analyze Grass Leaf Growth Using Function Fitting

Wannes Voorend[1,2,3], Peter Lootens[3], Hilde Nelissen[1,2], Isabel Roldán-Ruiz[3], Dirk Inzé[1,2] and Hilde Muylle[3*]

[1]Department of Plant Systems Biology, VIB, Technologiepark 927, 9052 Gent, Belgium

[2]Department of Plant Biotechnology and Bioinformatics, Ghent University, Technologiepark 927, 9052 Gent, Belgium

[3]Plant Sciences Unit – Growth and Development, Institute for Agricultural and Fisheries Research (ILVO), Caritasstraat 21, 9090 Melle, Belgium

# ABSTRACT

In grasses, leaf growth is often monitored to gain insights in growth processes, biomass accumulation, regrowth after cutting, etc. To study the growth dynamics of the grass leaf, its length is measured at regular time intervals to derive the leaf elongation rate (LER) profile over time. From the LER profile, parameters such as maximal LER and leaf elongation duration (LED), which are essential for detecting inter-genotype growth differences and/or quantifying plant growth responses to changing environmental conditions, can be determined. As growth is influenced by the circadian clock and, especially in grasses, changes in environmental conditions such as temperature and evaporative demand, the LER profiles show considerable experimental variation and thus often do not follow a smooth curve. Hence it is difficult to quantify the duration and timing of growth. For these reasons, the measured data points should be fitted using a suitable mathematical function, such as the beta sigmoid function for leaf elongation.

In the context of high-throughput phenotyping, we implemented the fitting of leaf growth measurements into a user-friendly Microsoft Excel-based macro, a tool called LEAF-E. LEAF-E allows to perform non-linear regression modeling of leaf length measurements suitable for robust and automated extraction of leaf growth parameters such as LER and LED from large datasets. LEAF-E is particularly useful to quantify the timing of leaf growth, which forms an important added value for detecting differences in leaf growth development. We illustrate the broad application range of LEAF-E using published and unpublished data sets of maize, *Miscanthus* spp. and *Brachypodium distachyon*, generated in independent experiments and for different purposes. In addition, we show that LEAF-E could also be used to fit datasets of other growth-related processes that follow the sigmoidal profile, such as cell length measurements along the leaf axis.

Given its user-friendliness, ability to quantify duration and timing of leaf growth and broad application range, LEAF-E is a tool that could be routinely used to study growth processes following the sigmoidal profile.

# BACKGROUND

Leaf growth has been monitored in a wide variety of grass species such as maize, rice, wheat, barley,*Lolium, Miscanthus, Sorghum* and *Brachypodium* making use of leaf length measurements taken at regular time intervals during development [1-8]. Parameters derived from these measurements such as leaf elongation rate (LER) and leaf elongation duration (LED) have been shown to be major determinants of individual and whole plant leaf area [9-14] and can be used to explain differences in final leaf length in response to environmental conditions and/or between genotypes [3,4,15].

In plant growth modeling, there is a growing consensus that approaches applying linear and exponential models are inadequate [16]. A linear fit assumes a constant LER over a longer period during leaf development [1,3,9,10] and an exponential or a log-linear relation assumes a constant relative elongation rate (RER). These assumptions limit the utility of the models, as both LER and RER may vary with environmental conditions and developmental stage [16]. The polynomial model does cope with variations in LER and RER during leaf development. However, polynomial functions tend to make spurious upward or downward predictions, especially at the extremes of the data [16,17]. Nonlinear regression is a more suitable strategy to describe leaf growth and to accommodate temporal variation in growth rates [16].

The beta sigmoid function, first used to describe whole plant growth [18], has been successfully applied to model the growth pattern of a single grass leaf [7,19]. Yin and coworkers [18] compared the performance of the beta sigmoid function with that of some other widely used sigmoid functions, such as Gompertz, Weibull and Richards to analyze datasets from maize, pea and wheat and concluded that the beta sigmoid function is unique in dealing with determinate growth [18]. This is due to the prediction of a zero growth rate at both the start and end of the determinate growth period which is characterized by three sub-phases: an early exponential growth phase, an approximately linear growth phase, followed by a steadily decelerating growth phase [20]. Furthermore, in contrast to other functions, the beta sigmoid function incorporates biologically relevant parameters and is highly flexible for describing various asymmetrical sigmoidal patterns [18].

In the context of high-throughput leaf phenotyping, there is a need for user-friendly tools that provide rapid and robust analysis of growth parameters from large datasets. Non-linear regression using function fitting is currently imbedded in statistical work packages such as SAS and R rendering the calculation, extraction and visualization of specific leaf growth parameters, such as LED, from large datasets difficult and time-consuming.

Here, we describe LEAF-E, a nonlinear regression-based tool for analyzing grass leaf growth data. The tool can be used to derive biologically relevant parameters such as final leaf length, maximal LER, LED but also parameters for the quantification of the timing of leaf growth, an important asset of this tool. To allow for the analysis of large datasets, the fitting procedure was automated in a user-friendly Microsoft Excel macro, which is innovative. We show how the application of this tool can assist data analysis and interpretation of experiments in which different genotypes or the response of single genotypes to different growth conditions are compared. For this purpose, we quantified and compared leaf growth parameters in published and unpublished datasets of three grass species: *Zea mays* (maize), *Brachypodium distachyon* and *Miscanthus spp.*

# RESULTS AND DISCUSSION

## Fitting of Kinematic Individual Leaf Length Measurements Using the Beta Sigmoid Function

### Goodness of Fit of the Beta Sigmoid Function for Maize, *Brachypodium*and *Miscanthus* Leaf Growth

First, we investigated to what extent the beta sigmoid function can be used to accurately fit leaf length measurements in function of thermal time or growing degree days (°Cd) in the three species considered here. Thermal time was used since plant growth and development are more closely related to accumulated mean daily temperature above a base value in the absence of other limiting conditions [21,22]. Equation 1 was used to fit length measurements of the 4th leaf over thermal time of nine B104 maize plants (dataset 1a, Figure 1). This resulted in $R^2$ values ranging from 0.9970 to 0.9989 with a mean value of 0.9981. Function fitting of leaf length measurements in *Miscanthus* and*Brachypodium* (datasets 2 and 3, respectively) rendered similar results: an overall mean $R^2$-value of 0.9931, ranging from 0.9669 to 0.9989 (n = 18) for the two *Miscanthus* species, and an overall mean $R^2$-value of 0.9932, ranging from 0.9871 to 0.9993 (n = 36) for the four *Brachypodium* inbred lines.

Plots of the fittings and $R^2$-values of individual plants of all datasets can be found in Additional file 1. A linear regression analysis of the measured leaf lengths versus the estimated value for those respective points in thermal time resulted in an $R^2$ value of 0.9986 for maize (dataset 1a), 0.9951 for*Miscanthus* (dataset 2) and 0.9940 for *Brachypodium*, indicating a good fit of the data for the three species investigated (Additional file 2 A, D and G). Nevertheless, analysis of the residuals showed a remaining sinusoidal pattern which indicates that part of the data could not be explained by the model (Additional file 3).

The higher $R^2$ value for the fittings of the maize dataset over the *Miscanthus*and *Brachypodium* datasets might be due to the more controlled environment of the growth chamber for maize as compared to the greenhouse for *Miscanthus* and *Brachypodium*. This indicates the importance of conducting growth experiments in a controlled environment and/or close monitoring of important growth regulating factors such as temperature, preferably at the level of the plant apex.

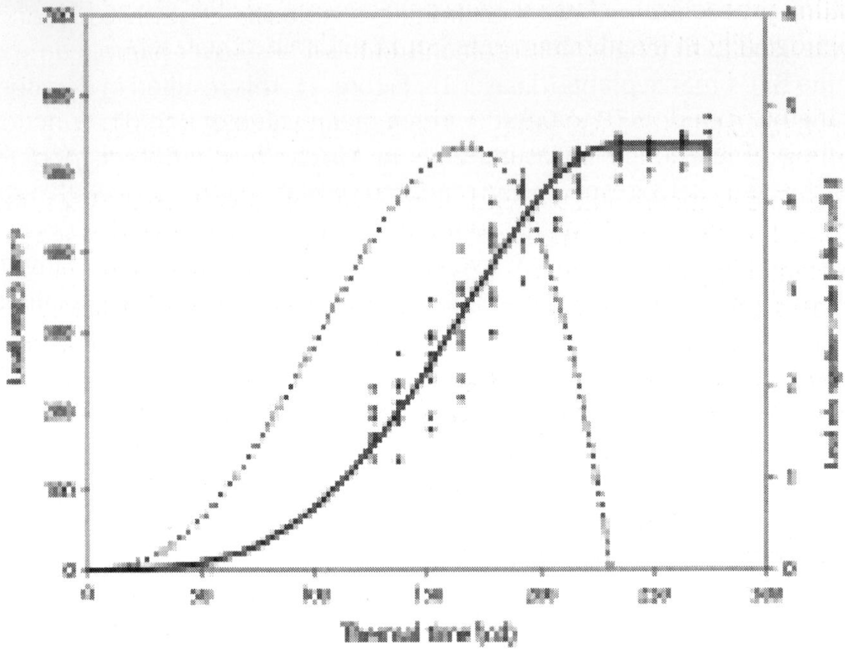

**Figure 1:** Leaf length and LER of maize B104 non-transgenic plants.Triangles represent length measurements of the 4th leaf of nine non-transgenic maize plants from dataset1a. The measurements of each individual leaf were fitted and biologically relevant growth parameters were extracted with LEAF-E. The S-shaped leaf length curve and bell-shaped LER curve are function plots using the mean values of the function parameters for the nine non-transgenic maize plants.

Based on these results, we can conclude that the beta sigmoid function is able to accommodate leaf growth measurements of three grass species with very distinct phenotypic characteristics. Maize and*Miscanthus spp.* both possess a 'C4' metabolism, however, maize is an annual crop characterized by one stem, whereas *Miscanthus spp.* are rhizomatous perennials that form numerous tillers.*Brachypodium* is a small, annual 'C3' plant used as a model for several temperate grain crops such as wheat and barley [23]. Based upon these findings and the results obtained previously in *L. perenne*[7], we can conclude that the beta sigmoid function is probably of broad application for describing leaf growth in both C3 and C4 grass species. For that reason, the beta sigmoid function was further used to develop an automated fitting procedure in an excel macro that we called LEAF-E and to derive biologically relevant parameters from the different datasets.

## Biologically Relevant Function Parameters

The Excel macro that we designed, automatically generates the fitted function parameters (the final leaf length *Lm*, the moment of maximal leaf elongation rate *tm*, and the moment at which leaf growth ceases *te*) and all additional parameters in the form of a table. For visual inspection, it also generates a graph showing the original data points, the fitted growth curve and the function parameters for each biological replicate (see Additional files 4 and 5). Although the tool can be used to fit the measurements of several replicates jointly, fitting data of individual leaves allows statistical analysis, such as estimation of averages and standard deviation values, for all growth parameters, and comparison of genotypes or treatments, a strategy that is both straightforward and statistically correct[17]. The advantage of using the beta sigmoid function in the form of Auzanneau and coworkers [7] is that the function parameters (*Lm*, *tm* and *te*) themselves are biologically relevant when assessing grass leaf growth. This represents a clear advantage over functions that are based on parameters with no biologically relevant meaning or parameters that are difficult to interpret visually such as the Weibull equation [18]. In addition, parameters that allow for the quantification of the timing of leaf growth, such as *tm* and *te*, can be estimated, which is not trivial using other methodologies such as a log-linear, polynomial or linear fit.

We used the nine B104 maize plants from dataset 1a to illustrate how leaf growth can be analyzed using LEAF-E (Figure 1, Table 1). The final length (*Lm*) of the $4^{th}$ leaf in the nine B104 maize plants was 535 ± 6 mm on average. This value was attained after 231 ± 5°Cd (*te*), which, in this experiment, is equivalent to 16.5 days after sowing. The moment at which the LER was maximal, *tm*, was 167 ± 5°Cd, or 12.0 days after sowing.

**Table 1:** Effect of *GA20ox1* overexpression on maize leaf elongation

| Growth parameter | | | ***AtGA20ox1* OE (mean ± SE)** | **Control (mean ± SE)** | **Difference in mean(+)** | |
|---|---|---|---|---|---|---|
| | Lm | (mm) | 743 ± 13 | 535 ± 6 | 38.9% | *** |
| | LERmax | (mm°C$^{-1}$d$^{-1}$) | 6.2 ± 0.2 | 4.6 ± 0.1 | 34.2% | *** |
| Thermal time points | t100 | (°Cd) | 107 ± 2 | 108 ± 4 | -1.7% | NS |
| | t20% | (°Cd) | 121 ± 3 | 111 ± 4 | 8.7% | NS |
| | t50% | (°Cd) | 165 ± 3 | 153 ± 5 | 7.6% | * |
| | tm | (°Cd) | 180 ± 4 | 167 ± 5 | 7.7% | * |
| | t90% | (°Cd) | 217 ± 4 | 203 ± 5 | 6.7% | * |
| | te | (°Cd) | 246 ± 5 | 231 ± 5 | 6.3% | * |
| Leaf elongation durations | LED(100-e) | (°Cd) | 139 ± 4 | 123 ± 2 | 13.3% | ** |
| | LED(20%-90%) | (°Cd) | 96 ± 2 | 92 ± 2 | 4.2% | NS |
| | LED(20%-e) | (°Cd) | 125 ± 30 | 120 ± 2 | 4.0% | NS |

A segregating population produced by backcrossing (BC) a transgenic plant overexpressing the *Arabidopsis thaliana GIBBERELLIC ACID 20 OXIDASE1 (GA20ox1) gene* **to the wild-type line B104 was** analysed for leaf growth. The results are based on the analysis of eleven transgenic and nine non-transgenic BC1 plants. Lm: final leaf length; LERmax: maximal leaf elongation rate; t20%, t50%, t90%, te: time points at which the leaf reaches 20%, 50%, 90% and 100% of the final leaf length, respectively; t100: time point at which the leaf reaches 100 mm; tm: time point at which the leaf reaches LERmax; LEDs: leaf elongation durations between above stated thermal time points.

+Statistical significance based on student t-test of non-transgenic plants (n = 9) vs *GA20ox1* **overexpression (n** = 11), *$p < 0.05$, ** $p < 0.01$, ***$p < 0.001$, NS non-significant. Applied base temperature for thermal time calculation = 10°C, Mean of overall $R^2$ **values** = 0.9983 (0.9970-0.9991).

## Flexibility to Extract Additional Biologically Relevant Information from the Dataset

The major advantage of fitting a continuous function to the data is that for any given thermal time *t*(°Cd), the leaf length *L* (mm) can be estimated and vice versa (Figure 2).

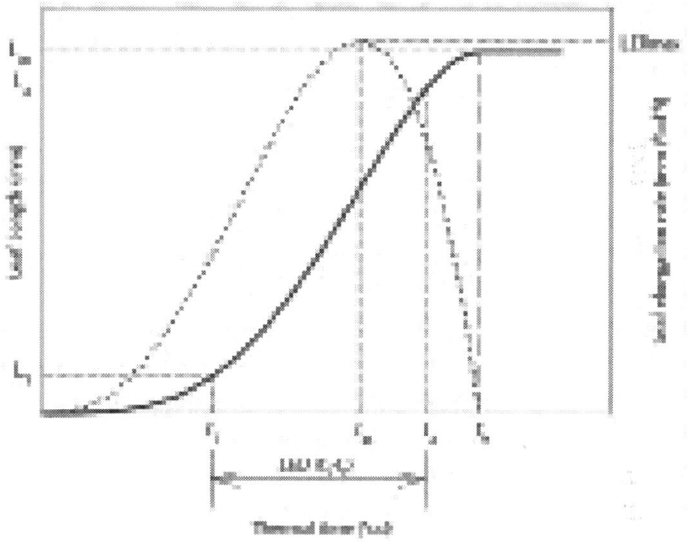

**Figure 2:** Deriving leaf growth parameters from the fitted leaf length and LER curve using LEAF-E. The leaf length curve (S-shaped) is generated by fitting Equation 1 to measurements of a single leaf. Based upon the leaf length curve the final leaf length *Lm* can be calculated as well as the thermal time *t* needed to reach any given leaf length L, e.g. *te* is the thermal time needed to reach *Lm*. As a result, LEDs can be calculated between every desirable pair of thermal time points, e.g. LED(t1-t2). The first derivative of Equation 1 renders a bell-shaped LER curve from which maximal leaf elongation rate *LERmax*, occurring at the thermal time point *tm*, can be extracted.

For example, the leaves of the nine B104 maize plants of dataset 1a (Table 1) reached 100 mm (*t100*) at 108 ± 4°Cd. In this experiment this was just before the 4$^{th}$ leaf emerged from the pseudo-stem or whorl (the spiral arrangement of leaves forming a cylindrical structure from where newly formed leaves emerge), approximately 8 days after sowing.

Knowing that the average final length (*Lm*) is 535 ± 6 mm, we can state that a considerable share of at least 19% of the final maize leaf length is hidden in the pseudo-stem of this genotype. For *t20%*, the moment at which the 4th leaf reaches 20% of its final length, we obtained an estimate of 111 ± 4°Cd or 7.9 days after sowing. Exactly 50% of the final leaf length was attained at on average 153 ± 5°Cd (*t50%*), which is not significantly different from the moment of the maximal LER, *tm* at 167 ± 5°Cd. The leaf reached 90% of its final leaf length (*t90%*) at 203 ± 5°Cd or 14.5 days after sowing. The use of parameters that are independent of final leaf length, such as *t20%*, *t50%* and *t90%*, can however be meaningful when comparing genotypes that differ inherently in final leaf lengths. Likewise, we calculated a time window between reaching *20%* and *90%* of the final leaf length. The parameter was named [*LED(20%-90%)*] and corresponded to 92 ± 2°Cd or 6.7 days in this experiment (dataset 1a). Depending on the objective of a particular experiment, several other useful parameters can be extracted from the growth curve. Parameters can be deduced from the model, even at time points before empirical evidence can be taken (for maize e.g. t100 and in some cases also t20%) using non-destructive measurements of leaf length over time, as the leaf is then still hidden in the pseudo-stem. However, these are approximations and will need validation using destructive measurements of leaf elongation when one wants to investigate early leaf development, i.e. when the leaf is still hidden in the pseudo-stem [24,25].

## Leaf Elongation Rate and Steady-State Growth

Similarly, for any given thermal time *t* (°Cd), the LER (mm/°Cd) can be estimated using the first derivative of the beta sigmoid function (Equation 2). The maximum of this bell-shaped curve, is denoted as the maximal LER or *LERmax* (Figure 2). For the nine B104 maize plants (dataset 1a), we estimated a *LERmax* of 4.6 ± 0.1 mm/°Cd, equivalent to an impressive 2.7 mm/h or 64 mm/day.

Often, the LER profile is derived from calculations of leaf length increases between consecutive measurements divided by the respective thermal time interval. These calculations were performed on a plant-by-plant basis and can be viewed in Additional file 1. The LERmax is then defined as the maximal value in that LER profile. A linear regression analysis of the estimated LER and LERmax to the calculated LER and LERmax, respectively showed high $R^2$ values (Additional file 2), indicating a linear relationship between values obtained by the two approaches. The $R^2$ values of the linear regression were highest for the maize dataset (dataset 1a), 0.9067 and 0.9228 for LER and LERmax respectively (see Additional file 2 for $R^2$ values of all linear regression analyses).

Several studies have assumed that during the growth of a grass leaf a period of constant LER can be defined [5,15,25,26]. This assumption of steady-state growth in grass species such as maize has often been used as an acceptable simplification of the actual growth process [5,15,25,26], although it has been argued that this steady-state growth period is relatively short compared to the total leaf growth period [8,25,27]. LEAF-E allows to quantify a steady state growth period. This can be defined as a period in leaf development for which the LER is relatively stable, for example the thermal time window between the points at which LER has a value of 90% or 95% of LERmax. For the maize B104 plants of dataset 1a, the thermal time window for which the LER was higher than 90% and 95% of LERmax was 34.5 ± 0.6°Cd or 2.5 days (from 10.7 until 13.1 days after sowing) and 49.1 ± 0.9°Cd or 3.5 days (from 10.1 until 13.6 days after sowing), respectively. We therefore conclude that, for the investigated dataset, a relatively stable LER is found for a time-span of 2.5 to 3.5 days during leaf growth.

## Effect of *GA20ox1* Overexpression on Maize Leaf Elongation

Comparison of transgenic maize plants overexpressing the GA biosynthesis gene *GA20ox1* with non-transgenic plants in a previous study, demonstrated that altering GA levels specifically affects the size of the division zone, resulting in proportional changes in leaf and whole plant growth rates [28]. This published dataset was used to validate the LEAF-E tool.

Overexpression of *GA20ox1* in maize resulted into significantly longer leaves (*Lm*) and higher maximum leaf elongation rates (*LERmax*) (39% and 34% respectively, $p < 0.001$). This corresponded very well with the 38% increase in LER reported earlier by Nelissen and coworkers [28]. However, in addition to the calculation of *Lm* and *LERmax*,

LEAF-E also facilitates analysis of parameters that describe the timing of leaf growth. For example, leaves of transgenic plants took slightly, but significantly more time to reach their full length (*te*) than those of non-transgenic plants (Figure 3A and Table 1). As a consequence, also *t50%*, *t90%* and *te*, as well as *tm*, occurred significantly later in the transgenics. This means that leaves of transgenic plants took slightly more time to reach 50%, 90% and 100% of their final leaf length.

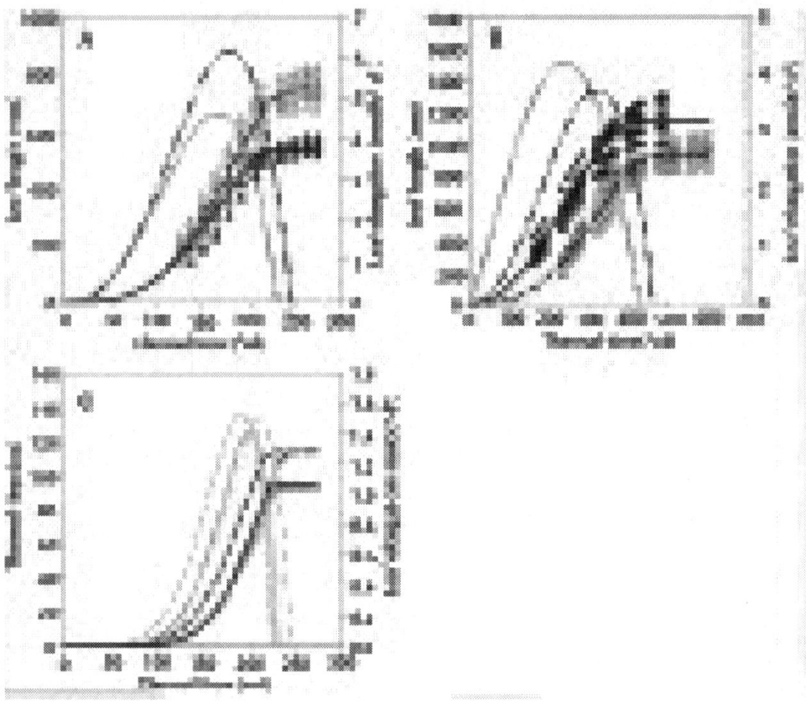

**Figure 3:** Analysis of the leaf elongation datasets of maize,*Miscanthus* and *Brachypodium* using LEAF-E. (A) Leaf length measurements of transgenic (white triangle) and non-transgenic (black triangle) plants of a segregating population produced by backcrossing a transgenic plant overexpressing the *GA20ox1 gene* to the wild-type line B104 maize, including leaf length (s-shaped) and LER (bell-shaped) function plots for both groups using the mean values of the function parameters. (B) Leaf length measurements of*Miscanthus sinensis* 'Goliath' (black triangle) and *M. x giganteus* (white triangle) plants, including leaf length and LER function plots for both groups using the mean values of the function parameters. (C) Leaf length and LER function plots for *Brachypodium distachyon*inbred lines Bd3-1 (full line), Bd21 (dotted), Bd21-

3 (dashed) and Bd2-3 (dash dotted) using the mean values of the function parameters. For the sake of clarity, the individual leaf length measurements are not shown in this case.

Using LEAF-E to reanalyze dataset 1a, previous findings were confirmed but this analysis allowed a more detailed study of the timing of leaf growth, thereby facilitating the detection of dissimilarities such as a shift in attaining *LERmax* in *GA20ox1* overexpressing plants, that could not be quantified previously.

## Variation in Leaf Growth Behavior in Two *Miscanthus* Species

We investigated the leaf growth characteristics of two genotypes of *Miscanthus* belonging to different species with high potential as bio-energy crops, but with contrasting phenotypic characteristics. *M. sinensis 'Goliath'* is characterized by high shoot densities, whereas *M. x giganteus* produces less, but thicker and taller shoots [29,30].

When comparing leaf growth characteristics of *M. sinensis 'Goliath'* and *M. x giganteus*, we found that *M. sinensis 'Goliath'* had significantly longer leaves than *M. x giganteus* (Table 2, Figure 3B). The leaves of *M. sinensis 'Goliath'* on average attained a length of 1140 ± 46 mm, whereas leaves of*M. x giganteus* on average became 923 ± 26 mm in length (Table 2). The longer leaves of *M. sinensis 'Goliath'* plants cannot be explained by significant changes in *LERmax* nor by an extended elongation period (no significant differences for LED values). However, our analysis revealed that leaves of *M. sinensis 'Goliath'* plants display a very strong initial growth compared to *M. x giganteus*. Parameters *t20%*, *t50%* and *tm* are all attained sooner (Table 2). Leaves of *M. sinensis 'Goliath'*reached 50% of their final leaf length at 230 ± 8°Cd, or 64°Cd sooner than *M. x giganteus* plants, at which point the leaves of the last were just emerging from the whorl.

**Table 2:** Comparison of leaf elongation in two *Miscanthus* genotypes from different species

| **Growth parameter** | | | ***M. Sinensis* 'Goliath' (mean ± SE)** | ***M. x giganteus* (mean ± SE)** | **Difference in mean(+)** | |
|---|---|---|---|---|---|---|
| | Lm | (mm) | 1140 ± 46 | 923 ± 26 | -217 | ** |
| | LERmax | ($mm°C^{-1}d^{-1}$) | 4.2 ± 0.2 | 3.8 ± 0.2 | -0.4 | NS |
| Thermal time points | t100 | (°Cd) | 95 ± 4 | 167 ± 4 | 71 | *** |
| | t20% | (°Cd) | 141 ± 7 | 207 ± 5 | 65 | *** |
| | t50% | (°Cd) | 230 ± 8 | 294 ± 8 | 64 | *** |
| | tm | (°Cd) | 240 ± 10 | 320 ± 8 | 80 | *** |
| | t90% | (°Cd) | 350 ± 10 | 400 ± 13.8 | 50 | NS |
| | te | (°Cd) | 425 ± 10 | 461 ± 18 | 37 | NS |
| Leaf elongation duration | LED(100-e) | (°Cd) | 329 ± 8 | 295 ± 18 | -35 | NS |
| | LED(20%-90%) | (°Cd) | 209 ± 5 | 194 ± 11 | -15 | NS |
| | LED(20%-e) | (°Cd) | 283 ± 7 | 255 ± 16 | -29 | NS |

The results are based on the analysis of nine *M. sinensis* 'Goliath' and eight *M.* x *giganteus* plants. Lm: final leaf length; LERmax: maximal leaf elongation rate; t20%, t50%, t90%, te: time points at which the leaf reaches 20%, 50%, 90% and 100% of the final leaf length, respectively; tm: time point at which the leaf reaches LERmax; LEDs: leaf elongation durations between above stated thermal time points.

+Statistical significance based on student t-test of *M. sinensis* 'Goliath' ($n=9$) vs *M.* x *giganteus* ($n=8$), **$p<0.01$, ***$p<0.001$, NS non-significant. Applied base temperature for thermal time calculation = 8°C, Mean of overall $R^2$ values = 0.9932 (0.9695-0.9989).

The analysis with LEAF-E shows that the leaf growth pattern in these two *Miscanthus spp.* is significantly different and that the differences in total leaf length can be attributed to contrasting early leaf growth.

## Variation in Leaf Growth Behavior in Different *Brachypodium Distachyon* Inbred Lines

Fifty *Brachypodium distachyon* inbred lines are currently being used for a study of natural diversity, which is led by 'The International Brachypodium Initiative' [23]. We analyzed the leaf growth behavior of four diploid *Brachypodium* inbred lines that are part of that study: Bd21, Bd21-3, Bd2-3 and Bd3-1 (Figure 3C).

Leaf growth analysis of these four genotypes with LEAF-E revealed distinct leaf growth characteristics. Based upon final leaf length, two groups can be distinguished. Bd21 and Bd3-1 have short leaves and Bd21-3 and Bd2-3 have long leaves (Figure 3C, Table 3). The length of the leaves is determined by both LER and LED. For Bd21, a low LER is probably the underlying factor of the shorter leaves (Table 3). This is in contrast to Bd3-1 which, like Bd21, has short leaves but a*LERmax* that is similar to those of the genotypes with longer leaves (Table 3). Parameters for LED are the smallest for Bd3-1. Thus, the leaf of Bd3-1 is short, most likely due to a short growing period. Bd21-3 and Bd2-3 both have long leaves, a high *LERmax* and similar LED. Despite these similarities, based upon our analysis, we can conclude that the 3$^{rd}$ leaf of Bd21-3 plants starts and finishes its growth significantly earlier in thermal time than that of Bd2-3 (Table 3). These results suggest that different mechanisms might drive leaf growth in different accessions. In *Arabidopsis* it was found that at least five different mechanisms contribute to final leaf size [31].

**Table 3:** Comparison of leaf elongation in four *Brachypodium* inbred lines

| Growth parameter | | | Bd3-1 (mean ± SE) (+) | Bd21 (mean ± SE) (+) | Bd21-3 (mean ± SE) (+) | Bd2-3 (mean ± SE) (+) |
|---|---|---|---|---|---|---|
| | Lm | (mm) | $94^{a} \pm 3$ | $91^{a} \pm 1$ | $113^{b} \pm 3$ | $113^{b} \pm 2$ |
| | LERmax | ($mm°C^{-1}d^{-1}$) | $1.44^{ab} \pm 0.05$ | $1.29^{a} \pm 0.03$ | $1.52^{b} \pm 0.05$ | $1.51^{b} \pm 0.03$ |
| Thermal time points | t20% | (°Cd) | $162\ ^{a} \pm 3$ | $163^{a} \pm 2$ | $148^{b} \pm 3$ | $166^{a} \pm 2$ |
| | t20 | (°Cd) | $164^{a} \pm 3$ | $166^{a} \pm 2$ | $145^{b} \pm 3$ | $163^{a} \pm 2$ |
| | t50% | (°Cd) | $189^{a} \pm 2$ | $193^{a} \pm 2$ | $178^{b} \pm 3$ | $197^{a} \pm 2$ |
| | tm | (°Cd) | $202^{a} \pm 2$ | $206^{a} \pm 2$ | $192^{b} \pm 3$ | $211^{a} \pm 2$ |
| | t90% | (°Cd) | $217^{ab} \pm 1$ | $222^{bc} \pm 2$ | $210^{a} \pm 4$ | $229^{c} \pm 2$ |
| | te | (°Cd) | $230.8^{ab} \pm 0.8$ | $238^{bc} \pm 2$ | $226^{a} \pm 4$ | $245^{c} \pm 2$ |
| Leaf elongation durations | LED(20-e) | (°Cd) | $67^{a} \pm 3$ | $72^{a} \pm 1$ | $81^{b} \pm 2$ | $82^{b} \pm 2$ |
| | LED(20%-90%) | (°Cd) | $69^{a} \pm 2$ | $75^{ab} \pm 1$ | $78^{b} \pm 1$ | $79^{b} \pm 1$ |
| | LED(20%-e) | (°Cd) | $55^{a} \pm 2$ | $59^{ab} \pm 1$ | $62^{b} \pm 1$ | $63^{b} \pm 1$ |

Bd3-1 (n = 7), Bd21 (n = 10) plants, Bd21-3 (n = 7) plants, Bd2-3 (n = 10). Lm: final leaf length; LERmax: maximal leaf elongation rate; t20%, t50%, t90%, te: time points at which the leaf reaches 20%, 50%, 90% and 100% of the final leaf length, respectively; tm: time point at which the leaf reaches LERmax; LEDs: leaf elongation durations between above stated thermal time points.

+Statistical significance indicated with distinct letters based on ANOVA and Scheffé Post hoc test ($p < 0.05$) between lines Bd3-1 (n = 7), Bd21 (n = 10) plants, Bd21-3 (n = 7) plants, Bd2-3 (n = 10), applied base temperature for thermal time calculation = 11 °C, Mean of overall $R^2$ values = 0.9993 (0.9871-0.9993).

## Fitting of Cell Length Measurements along the Leaf Axis of Maize Overexpressing *Ga20ox1* Using the Beta Sigmoid Function

The cell length profile along the longitudinal axis of an actively growing grass leaf also displays a sigmoid pattern [8]. This sigmoidal profile is determined by the spatial distribution of cells in different stages of differentiation along the leaf axis: a number of dividing cells of small size at the leaf base, a stretch of cells that undergo elongation and thus increase in length when being pushed towards the leaf tip, and finally the tip of the leaf that is made up of cells that have reached their final length. We applied LEAF-E, adapted to use the extended version of the beta sigmoid function (Equation 3), to fit cell length measurements of dataset 1b. We found that fitting was successful and resulted in overall $R^2$-values ranging from 0.8420 up to 0.8749 for cell length measurements of the*GA20ox1* overexpressing and control plants. Knowing that cell lengths can vary considerably, even for adjacent cells of the same cell file in one leaf (Figure 4), these $R^2$-values are noticeably high. However, a goodness of fit assessment on cell length measurements is, to our knowledge, not reported in literature, making any comparison difficult. We explored the suitability of three other commonly used sigmoidal functions, namely the Weibull, the Logistic and the Gompertz functions for fitting cell length data of the control plants. The fits of cell length data resulted in $R^2$-values of 0.8504 ± 0.0054 for Weibull, 0.8485 ± 0.0055 for logistics, 0.8499 ± 0.0053 for beta sigmoid and 0.8495 ± 0.0054 for Gompertz. The goodness of fit was also evaluated by linear regression of the estimated to the measured values (Additional file 6) and again found highly similar $R^2$-values. For the beta sigmoid function, we investigated the deviation of the data points from the fitted curve and could detect a slight underestimation of cell lengths in the very first 3 mm from the leaf base (Additional file 3D). This is because the cells in the first few millimeter of the division zone are slightly longer than those in the middle of the division zone; this hyperbolic profile at the leaf basis cannot be fitted with the functions that we used here, since they assume an increment in cell lengths. Additionally, the value of the residuals increased with the distance from the leaf base (Additional files3D and 6). For this dataset, the use of the beta sigmoid function in the form of Equation 3 is preferred over the other investigated functions because the function parameters themselves provide relevant biological information. However, the exploration of mathematical functions that are more appropriate for fitting cell lengths along the longitudinal axis of grass leaves should be the topic of future research.

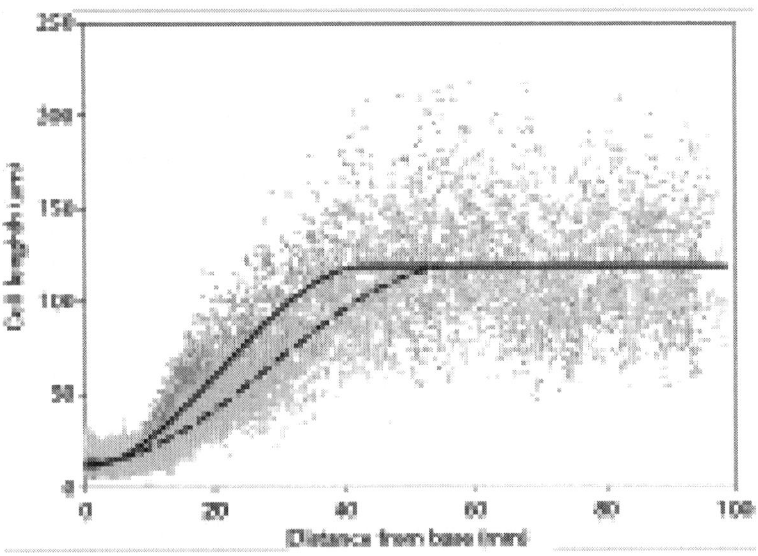

**Figure 4:** Effect of *GA20ox1* overexpression on the cell length profile of the 4th leaf in maize. The cell length profile along the axis of the 4th leaf is shown for three non-transgenic (triangle, dark gray) and three transgenic (x, light gray) plants. The S-shaped curves are function plots using the mean values of the fitted function parameters for non-transgenic (full) and transgenic (dashed) profiles.

The cell length profile along the leaf axis in maize plants overexpressing *GA20ox1* (dataset 1b) was analyzed by Nelissen and coworkers [28] using a polynomial fit approach similar to Rymen and coworkers [3]. Using LEAF-E, we determined that the size of the division zone, defined as the stretch of cells near the leaf base by which function fitted cell lengths do not exceed 40 μm (obtained by DAPI-staining of cells along the proximal-distal axis), is on average 33% longer ($p < 0.01$) in*GA20ox1* overexpressing leaves as compared to non-transgenic leaves (Figure 4, Table 4). We estimated that the elongation zone, defined as the distance between the end of the division zone and the function parameter *Pe* (position in the leaf where cells reach their maximal length, equivalent to*te* for leaf elongation), is 29% longer ($p < 0.01$) in *GA20ox1* overexpressing leaves. Moreover, we can state that the position where maximal cell elongation occurs (*Pm*, equivalent to *tm* for leaf elongation) is situated 8 mm (or 38%) further away from the leaf base in *GA20ox1* overexpressing leaves as compared to non-transgenic plants and that the mature cell length *Lm* (mm) did not differ ($p = 0.92$). These findings are in accordance with results obtained earlier by Nelissen and coworkers [28]. The analysis using LEAF-E thus allowed a more detailed study of the cell length profile, which facilitated the detection of a longer elongation zone in *GA20ox1* overexpressing plants and a shift in the position of maximal cell elongation, that could not be quantified previously. Nevertheless, it might still be necessary to validate the estimation of the size of the division zone using DAPI staining or a similar method, as was performed here. We believe that function fitting, using LEAF-E, can be very useful and might even be a necessary step for the analysis of cell length profiles in the grass leaf in future experiments.

**Table 4:** Effect of GA20ox1 overexpression on the maize cell length profile

| Parameter | | | *AtGA20ox1* OE (mean ± SE) | Control (mean ± SE) | Difference in mean(+) | |
|---|---|---|---|---|---|---|
| Cell length | Lb | (μm) | 13.2 ± 1.7 | 10.9 ± 0.3 | 22% | NS |
| | Lm | (μm) | 118.1 ± 0.6 | 118.5 ± 3.8 | 0% | NS |
| Position along the leaf axis | Pm | (mm) | 29.5 ± 2.1 | 21.3 ± 0.8 | 38% | ** |
| | Pe | (mm) | 56.6 ± 2.5 | 43.4 ± 0.3 | -30% | ** |
| Zones along the leaf axis | division zone (cell < =40 μm) | (mm) | 19.5 ± 0.5 | 14.6 ± 0.3 | 33% | ** |
| | elongation zone (cell > 40 μm - Pe) | (mm) | 37.1 ± 1.0 | 28.7 ± 0.9 | -29% | ** |

A segregating population produced by backcrossing (BC) a transgenic plant overexpressing the *Arabidopsis thaliana GIBBERELLIC ACID 20 OXIDASE1 (GA20ox1)* gene to the wild-type line B104 was analyzed for the cell lengths profile along the leaf axis. The results are based on the analysis of three transgenic and three non-transgenic BC1 plants. Lb: initial cell length Lm: final cell length; Pm: position along the leaf axis with maximal cell elongation rate; Pe: position along the leaf axis where cells reach their final length.

+Statistical significance based on student t-test of non-transgenic plants ($n = 3$) vs *GA20ox1* overexpression ($n = 3$), $^{**}p < 0.01$, NS non-significant.

# CONCLUSIONS

Here, we provide a tool that we named LEAF-E for fast and straightforward analysis of grass leaf growth data using nonlinear regression in a simple Microsoft Excel format. We automated the fitting procedure using Excel 2010 and the Solver function (32 bit). The results of fitting the beta sigmoid function to the leaf length measurements are stored in tabular form and can easily be analyzed in standard statistical software programs in search of differences due to the applied treatments or to explore inter-genotypic or inter-population differences.

We applied LEAF-E to three datasets containing leaf length measurements of maize, *Miscanthus* and*Brachypodium*. In *Miscanthus* and *Brachypodium*, we have shown that LEAF-E is an appropriate tool for data analysis and that the analyzed species and genotypes display distinct leaf growth characteristics. In maize, the changes in both leaf elongation and cell length profile along the leaf axis as a result of enhanced GA levels, previously demonstrated by Nelissen and coworkers [28], were confirmed. In addition, we demonstrated that using LEAF-E, the timing of leaf growth can be studied, thereby facilitating the detection of dissimilarities in the timing of leaf growth that could not be quantified using other approaches. Furthermore, analysis with LEAF-E allows for a robust calculation of *LERmax* and LED, which leads to reliable detection of significant changes. Therefore, we propose LEAF-E as an excellent tool for comparing leaf growth behavior in different genotypes or to analyze the response of specific genotypes to a treatment.

# METHODS

## Datasets Used For Validation

### Dataset 1a

The dataset on leaf elongation of the maize B104 inbred line overexpressing the *Arabidopsis thaliana GIBBERELLIC ACID 20 OXIDASE1* (*GA20ox1*) gene, previously described by Nelissen and coworkers[28] was reanalyzed here. A segregating population produced by backcrossing the overexpression line (hemizygous for the transgenic event) to the wild-type B104 inbred line and consisting of 9 non-transgenic and 11 transgenic plants was used.

To determine leaf elongation rates, the length of the $4^{th}$ leaf of transgenic and non-transgenic plants was measured daily until complete development, as previously described [28]. The plants were grown in a growth chamber at 24°C. Here we used a base temperature of 10°C for thermal time (Growing degree days, GDD) calculations. For further details about growth conditions see [28].

## Dataset 1b

The same segregating population, used to generate dataset 1a, was previously used by Nelissen and coworkers [28] for the analysis of cell lengths along the leaf axis, based upon methods previously described [32]. In short, the 4$^{th}$ leaf was harvested two days after appearance from the pseudo-stem (stem-like structure composed of concentric rolled or folded blades and sheaths that surround the growing point). At this time point, the ligule is only a few mm away from the base of the plant. The length of cell files adjacent to stomatal rows along the proximal-distal axis was measured using a DIC microscope (AxioImager, Zeiss, USA), and image analysis software (AxioVision, Zeiss, USA). The size of the division zone was determined as the distance between the base and the most distally observed mitotic figure in DAPI-stained leaves along the proximal-distal axis, with a fluorescence microscope (AxioImager, Zeiss, USA). Here we reanalyzed the cell length measurements.

## Dataset 2

Eight *Miscanthus* x *giganteus* and nine *M. sinensis* 'Goliath' plants were grown at 20°C (average temperature over the measuring period was 19.1°C) in a greenhouse, in Melle, Belgium, in September 2012 with no supplementary light. Plants were grown from rhizome cuttings in 2-l pots and were hand-watered and not fertilized during the experiment. The rhizomes were excavated during the winter of 2011 and stored in a cold room at 3°C until the start of the experiment. The length of the 4$^{th}$ leaf was measured five times a week (from leaf tip to soil level). Based on our observations at this developmental stage, the contribution of internode elongation to the height measurements as leaf lengths can be neglected and transition to flowering has not yet occurred. The measurements were spread over a time period of approximately four weeks. The calculation of thermal time was based on the average air temperature in the greenhouse taking into account a base temperature of 8°C, based on Farrell and coworkers [33].

### Dataset 3

The *Brachypodium distachyon* inbred lines Bd21, Bd2-3 and Bd3-1 were provided by David F. Garvin from the USDA-ARS (Minnesota, US), and line Bd21-3 was provided by Richard Sibout from INRA-IJPB (Versailles, France). Plants were grown in rootrainers (Haxnicks®, UK) in biological replicates (n = 10, 10, 7, 9 respectively) in a greenhouse at an average temperature of 21°C in Melle, Belgium, August 2012 with no supplementary light. To calculate the thermal time, a base temperature of 10°C was used. Fertilizer was added with the water supply: conductivity Ec = 1mS/cm; water soluble fertilizer Poly-feed (Haifa, Belgium) (N, $P_2O_5$, $K_20$; 20:5:20 + 3 MgO). Measurements were taken from the tip of the 3rd leaf to its basal level on a daily basis, for a period of 10 days. Based on our observations at this developmental stage, the contribution of internode elongation to the height measurements as leaf lengths can be neglected and transition to flowering has not yet occurred.

## A Mathematical Function for Fitting Leaf Length Measurements

For the estimation of leaf growth parameters we used the beta sigmoid function for determinate growth, inspired by the Euler integral, in the form of Equation 1. This function was used previously by Auzanneau and coworkers [7] and Verdenal and coworkers [19] to model leaf growth after cutting in *Lolium perenne*. In short, the leaf length *L* (mm) at a given moment in development *t* (°Cd) is determined by final leaf length *Lm* (mm) and three particular points in leaf development, expressed as units of thermal time or growing degree days (°Cd). Thermal time is a summation of cumulative differences between daily mean temperature and a specified base temperature, below which the plant does not grow or grows very slowly [22]. These thermal time points are the moment at which leaf growth starts *t0* (°Cd), the moment of maximal leaf growth rate *tm* (°Cd) and the moment at which leaf growth ceases *te* (°Cd). Estimations of *t0* often result in negative values that are biologically not relevant [7]. Therefore, in the experiments in which seedlings were involved (maize and*Brachypodium*), we assumed that *t0* = 0 was at the moment of sowing. In the case of *Miscanthus*, $t0 = 0$ was assumed to be at the moment of potting the rhizomes (no visible leaves at this stage).

$$L = Lm.\left(1+\frac{(te-t)}{(te-tm)}\right).\left(\frac{(t-t0)}{(te-t0)}\right)^{\frac{(te-t0)}{(te-tm)}} \qquad (1)$$

Equation 1. Beta sigmoid function for fitting leaf length, modified from [7]. Function is applicable for $t0 \leq t \leq te$ and $t0 \leq tm < te$. For $t > te$, Equation 1 is reduced to $L = Lm$.

The leaf elongation rate (LER) at any given moment in leaf development *t* (°Cd) can be calculated from the LER function (Equation 2), which is the first derivative of Equation 1. From this equation we determined the maximum leaf elongation rate or *LERmax* (mm/°Cd), as the LER at *tm*.

$$\frac{dL}{dt} = Lm.\left(\frac{\left(1+\frac{(te-t)}{(te-tm)}\right).\left(\frac{(t-t0)}{(te-t0)}\right)^{\left(\frac{(te-t0)}{(te-tm)}-1\right)}}{(te-tm)} - \frac{\left(\frac{(t-t0)}{(te-tc)}\right)^{\frac{(te-to)}{(te-tm)}}}{(te-tm)}\right) \qquad (2)$$

Equation 2. Leaf elongation rate function, modified from [7].

As Equation 1 is a continuous function, it allows calculating the leaf length *L* (mm) at any given moment in the leaf elongation period *t*, and vice versa (Figure 2). Therefore, in addition to the parameters *Lm*, *t0*, *tm* and *te*, we estimated a set of parameters that can be biologically relevant. For the maize dataset 1a we estimated the time point *t100*, the moment at which the leaf length is 100 mm. The *t100* time point, which is early in development, was chosen since it is close to the moment at which the leaf emerges from the pseudo-stem in maize non-transgenic B104 plants. Furthermore, we estimated *t20%*, *t50%* and *t90%* (°Cd), which are the moments at which the leaf reaches 20%, 50% and 90% of its final length, respectively. For *Brachypodium* (dataset 2), we replaced the *t100* parameter by *t20* (°Cd), the moment at which the leaf reaches 20 mm in length, to accommodate the smaller size of the *Brachypodium* leaf. These extra parameters allow the comparison of responses to different treatments or detect inter-genotypic differences in leaf growth development.

Expressing leaf growth as durations of thermal time, the leaf elongation duration or LED, allows describing leaf growth in a fluctuating environment [15] On that account, various LEDs were explored. For example, *LED(20%-90%)* defines the leaf growth duration between reaching 20% and 90% of its final length. Parameters *LED(100-e)* and *LED(20%-e)* were defined similarly.

## A Mathematical Function for Fitting Cell Length Measurements

Often, analysis of leaf growth involves cell length measurements along the leaf axis, providing insight in the sizes of dividing, elongating and mature cells, and of the leaf zones encompassing these three cell types [3,8,9,15,24,25]. Therefore, we assessed the fitting of cell length measurements along the leaf axis in maize with the beta sigmoid function. However, since in this case cells in the division zone have an initial length before proceeding to the elongation phase and eventually toward mature cells, a different equation had to be used as Equations 1 and 2 assume zero values at the beginning of growth. Yin and coworkers [18] describe an extended version of the beta sigmoid function that allows taking into account the initial length of the cell, *Lb* (mm) (Equation 3). In this case the data points do not represent a time series, but a positional-series of cell lengths along the leaf axis, taking the leaf base as position zero. For ease of interpretation, the symbols of the variable *t* and parameters *tm* and*te* from the original equation have been converted into the positions *p* (mm) along the leaf axis, starting from the base towards the tip. The position at which maximal cell elongation occurs is denoted as *Pm* (mm), and the position at which the cells cease to elongate is denoted as *Pe* (mm) (see Equation 3).

$$L = Lb + (Lm - Lb).\left(1 + \frac{(Pe - p)}{(Pe - Pm)}\right).\left(\frac{p}{Pe}\right)^{\frac{Pe}{(Pe-Pm)}} \qquad (3)$$

Equation 3. Extended version of the beta sigmoid function modified from [18] to fit cell length measurements along the leaf axis.

## LEAF-E: Function Fitting Using a Microsoft Excel Spreadsheet and the Solver Function

The fitting of leaf growth data using the beta sigmoid function was performed using Excel 2010 and the Solver function (32 bit) according to Brown [34]. The automation of the procedure, as described below, in the form of a macro is innovative. Each row in the datasheet contains the data of one individual leaf (ordered in a time series), the starting values of the parameters of the model, and the formulae to extract the necessary statistical components for the calculation of the least square estimates following Neter and coworkers [35]. First, the macro checks for non-empty rows. When a non-empty row is found, the model is fitted to the data by minimizing the sum of squares of the errors iteratively, and changing the starting values of the parameters at each step. Per row (=leaf), all values described in the previous sections are calculated and stored in a tabular form for further statistical analysis. In addition, a graph showing the original data points, the fitted growth curve and the function parameters is generated automatically. This enables the evaluation of the correlation coefficient and visual interpretation of the goodness of fit. It also provides an easy way to check for miss fits or errors in the data. Miss fitting can occur when the Solver function fails to minimize the sum of squares of the errors using a particular set of starting values. Accordingly, the procedure can be repeated using more appropriate starting values. This way of working guarantees a fast analysis, robust estimation of growth parameters and provides a table in standardized format containing the resulting parameters and derived parameters per leaf. The output data can easily be analyzed in search of differential responses using standard statistical software, depending on the experimental setup. We gave this Excel tool the name LEAF-E. LEAF-E is included as additional file to this article (Additional files 4 and 5, with a user manual on the first sheet). Additional file 4 includes test data of*M. sinensis* 'Goliath' leaf growth and default starting values prior running the macro and Additional file 5 includes the results of the fitting procedure and graphs after running the automated fitting procedure.

## Statistical Analysis

The statistical analysis of the derived growth parameters comprised a student t-test on datasets containing only two genotypes (datasets 1a, 1b and 2) and an ANOVA followed by post hoc Scheffé tests for datasets containing more than two genotypes (dataset 3). All these analyses were carried out in the software package *STATISTICA* version 11 (Statsoft Inc., USA). For the analysis of the goodness of fit of cell length data of dataset 1b, the fitting of the beta sigmoid and three other commonly used growth functions (Weibull, Gompertz and Logistic) and linear regression of the predicted to the observed values was carried out in *SIGMAPLOT* version 12_5 (Systat Software Inc., USA).

# ACKNOWLEDGEMENTS

We thank the Agency for Innovation by Science and Technology in Flanders (IWT) for funding and Simon Fonteyne (ILVO, Belgium) for providing leaf measurements of *Miscanthus* spp. We also like to thank David Garvin (USDA-ARS, US) and Richard Sibout (INRA-IJPB, France) for providing Brachypodium seed stocks.

# REFERENCES

1. Fricke W, McDonald AJS, Mattson-Djos L: Why do leaves and leaf cells of N-limited barley elongate at reduced rates? *Planta* 1997, 202:522-530.
2. Munns R, James RA: Screening methods for salinity tolerance: a case study with tetraploid wheat. *Plant Soil* 2003, 253:201-218.
3. Rymen B, Fiorani F, Kartal F, Vandepoele K, Inzé D, Beemster GTS: Cold nights impair leaf growth and cell cycle progression in maize through transcriptional changes of cell cycle genes. *Plant Physiol* 2007, 143:1429-1438.
4. Clifton-Brown JC, Jones MB: The thermal response of leaf extension rate in genotypes of the $C_4$-grass *Miscanthus*: an important factor in determining the potential productivity of different genotypes. *J Exp Bot* 1997, 48:1573-1581.

5. Lafarge T, Tardieu F: A model co-ordinating the elongation of all leaves of a sorghum cultivar was applied to both Mediterranean and Sahelian conditions. *J Exp Bot* 2002, 53:715-725.
6. Verelst W, Bertolini E, De Bodt S, Vandepoele K, Demeulenaere M, Pè ME, Inzé D: Molecular and physiological analysis of growth-limiting drought stress in *Brachypodium distachyon* leaves. *Mol Plant* 2013, 6:311-322.
7. Auzanneau J, Huyghe C, Escobar-Gutiérrez AJ, Julier B, Gastal F, Barre P: Association study between the gibberellic acid insensitive gene and leaf length in a Lolium perenne L. synthetic variety. *BMC Plant Biol* 2011, 11:183.
8. Parent B, Conejero G, Tardieu F: Spatial and temporal analysis of non-steady elongation of rice leaves. *Plant Cell Environ* 2009, 32:1561-1572.
9. Fiorani F, Beemster GTS, Bultynck L, Lambers H: Can meristematic activity determine variation in leaf size and elongation rate among four *Poa* species? A kinematic study. *Plant Physiol* 2000, 124:845-855.
10. Arredondo JT, Schnyder H: Components of leaf elongation rate and their relationship to specific leaf area in contrasting grasses. *New Phytol* 2003, 158:305-314.
11. Bultynck L, Ter Steege MW, Schortemeyer M, Poot P, Lambers H: From individual leaf elongation to whole shoot leaf area expansion: a comparison of three *Aegilops* and two *Triticum* species. *Ann Bot* 2004, 94:99-108.
12. Reymond M, Muller B, Tardieu F: Dealing with the genotype x environment interaction via a modelling approach: a comparison of QTLs of maize leaf length or width with QTLs of model parameters. *J Exp Bot* 2004, 55:2461-2472.
13. Chenu K, Chapman SC, Hammer GL, McLean G, Ben Haj Salah H, Tardieu F: Short-term responses of leaf growth rate to water deficit scale up to whole-plant and crop levels: an integrated modelling approach in maize. *Plant Cell Environ* 2008, 31:378-391.
14. Dignat G, Welcker C, Sawkins M, Ribaut JM, Tardieu F: The growths of leaves, shoots, roots and reproductive organs partly share their genetic control in maize plants. *Plant Cell Environ* 2013, 36:1105-1119.

15. Tardieu F, Reymond M, Hamard P, Granier C, Muller B: Spatial distributions of expansion rate, cell division rate and cell size in maize leaves: a synthesis of the effects of soil water status, evaporative demand and temperature. *J Exp Bot* 2000, 51:1505-1514.
16. Paine CET, Marthews TR, Vogt DR, Purves D, Rees M, Hector A, Turnbull LA: How to fit nonlinear plant growth models and calculate growth rates: an update for ecologists. *Methods Ecol Evol* 2012, 3:245-256.
17. Motulsky HJ, Ransnas LA: Fitting curves to data using nonlinear regression: a practical and nonmathematical review. *FASEB J* 1987, 1:365-374
18. Yin X, Goudriaan J, Lantinga EA, Vos J, Spiertz HJ: A flexible sigmoid function of determinate growth. *Ann Bot* 2003, 91:361-371.
19. Verdenal A, Combes D, Escobar-Gutiérrez AJ: A study of ryegrass architecture as a self-regulated system, using functional-structural plant modelling. *Funct Plant Biol* 2008, 35:911-924.
20. Goudriaan J, Van Laar HH: *Modelling Potential Crop Growth Processes: Textbook with Exercises*. Dordrecht; Boston: Kluwer Academic Publishers; 1994.
21. Neild R, Newman J: Growing season Characteristics and Requirements in the Corn Belt. In *Natl Corn Handb 40*. West Lafayette, IN 47907: Purdue University Cooperative Extension Service; 1990.
22. Atwell B, Kriedemann P, Turnbull C: *Plants in Action: Adaptation in Nature, Performance in Cultivation*. MacMillan Education Australia Pty Ltd: Melbourne, Australia; 1999.
23. International Brachypodium Initiative: Genome sequencing and analysis of the model grass *Brachypodium distachyon*. *Nature* 2010, 463:763-768.
24. Ben-Haj-Salah H, Tardieu F: Temperature affects expansion rate of maize leaves without change in spatial distribution of cell length (analysis of the coordination between cell division and cell expansion). *Plant Physiol* 1995, 109:861-870.
25. Muller B, Reymond M, Tardieu F: The elongation rate at the base of a maize leaf shows an invariant pattern during both the steady-

state elongation and the establishment of the elongation zone. *J Exp Bot* 2001, 52:1259-1268.

26. Bouchabké O, Tardieu F, Simonneau T: Leaf growth and turgor in growing cells of maize (*Zea mays* L.) respond to evaporative demand under moderate irrigation but not in water-saturated soil. *Plant Cell Environ* 2006, 29:1138-1148.
27. Durand J-L, Schäufele R, Gastal F: Grass leaf elongation rate as a function of developmental stage and temperature: morphological analysis and modelling. *Ann Bot* 1999, 83:577-588.
28. Nelissen H, Rymen B, Jikumaru Y, Demuynck K, Van Lijsebettens M, Kamiya Y, Inze D, Beemster GT: A local maximum in gibberellin levels regulates maize leaf growth by spatial control of cell division. *Curr Biol* 2012, 22:1183-1187.
29. Greef JM, Deuter M: Syntaxonomy of *Miscanthus x giganteus* GREEF et DEU. *Angew Bot* 1993, 67:87-90.
30. Hodkinson TR, Renvoize SA, Chase MW: Systematics of *Miscanthus*. *Asp Appl Biol* 1997, 49:189-198.
31. Gonzalez N, Vanhaeren H, Inze D: Leaf size control: complex coordination of cell division and expansion. *Trends Plant Sci* 2012, 17:332-340.
32. Rymen B, Coppens F, Dhondt S, Fiorani F, Beemster GTS: Kinematic Analysis of Cell Division and Expansion. In *Plant Developmental Biology (Methods in Molecular Biology, Vol 655)*. Edited by Hennig L, Köhler C. New York: Humana Press; 2010:203-227.
33. Farrell AD, Clifton-Brown JC, Lewandowski I, Jones MB: Genotypic variation in cold tolerance influences the yield of *Miscanthus*. *Ann Appl Biol* 2006, 149:337-345.
34. Brown AM: A step-by-step guide to non-linear regression analysis of experimental data using a Microsoft Excel spreadsheet. *Comput Methods Programs Biomed* 2001, 65:191-200.
35. Neter J, Kutner M, Nachtsheim C, Wasserman W: *Applied Linear Statistical Models*. 4th edition. McGraw-Hill Education: Columbus; 1996.

Chapter 10

# Use of Diffusion Magnetic Resonance Imaging to Correlate the Developmental Changes in Grape Berry Tissue Structure With Water Diffusion Patterns

Ryan J Dean[1], Timothy Stait-Gardner[1], Simon J Clarke[2], Suzy Y Rogiers[2,3], Gabriele Bobek[4] and William S Price[1,2,4*]

[1]Nanoscale Organisation and Dynamics Group, University of Western Sydney, Penrith NSW 2751, Australia

[2]National Wine & Grape Industry Centre, Charles Sturt University, Locked Bag 588, Wagga Wagga, New South Wales 2678, Australia

[3]New South Wales Department of Primary Industries, Locked Bag 588, Wagga Wagga, New South Wales 2678, Australia

[4]School of Medicine, University of Western Sydney, Penrith, NSW 2751, Australia

# ABSTRACT


## Background

Over the course of grape berry development, the tissues of the berry undergo numerous morphological transformations in response to processes such as water and solute accumulation and cell division, growth and senescence. These transformations are expected to produce changes to the diffusion of water through these tissues detectable using diffusion magnetic resonance imaging (MRI). To assess this non-invasive technique diffusion was examined over the course of grape berry development, and in plant tissues with contrasting oil content.

# RESULTS

In this study, the fruit of *Vitis vinfera* L. cv. Semillon at seven different stages of berry development, from four weeks post-anthesis to over-ripe, were imaged using diffusion tensor and transverse relaxation MRI acquisition protocols. Variations in diffusive motion between these stages of development were then linked to known events in the morphological development of the grape berry. Within the inner mesocarp of the berry, preferential directions of diffusion became increasingly apparent as immature berries increased in size and then declined as berries progressed through the ripening and senescence phases. Transverse relaxation images showed radial striation patterns throughout the sub-tissue, initiating at the septum and vascular systems located at the centre of the berry, and terminating at the boundary between the inner and outer mesocarp. This study confirms that these radial patterns are due to bands of cells of alternating width that extend across the inner mesocarp. Preferential directions of diffusion were also noted in young grape seed nucelli prior to their dehydration. These observations point towards a strong association between patterns of diffusion within grape berries and the underlying tissue structures across berry development. A diffusion tensor image of a post-harvest olive demonstrated that the technique is applicable to tissues with high oil content.

# CONCLUSION

This study demonstrates that diffusion MRI is a powerful and information rich technique for probing the internal microstructure of plant tissues. It was shown that macroscopic diffusion anisotropy patterns correlate with the microstructure of the major pericarp tissues of cv. Semillon grape berries, and that changes in grape berry tissue structure during berry development can be observed.

## BACKGROUND

Magnetic resonance imaging (MRI) is a powerful tool for investigating biological systems; it is non-invasive, does not use ionising radiation and can draw on a large variety of image contrasts [1,2]. For example, in the field of viticulture, MRI has demonstrated its ability to produce informative proton ($^1$H) density and transverse magnetic relaxation images of the internal structure of the grape berry [3,4]. Despite the good spatial resolution of the images from the aforementioned studies, it is possible to extract further microstructural (i.e. beyond the resolution limit of an MRI spectrometer) information from a grape berry. This can be achieved using diffusion MRI techniques to observe the distribution of diffusing molecules as they probe the edges of grape berry tissue microstructures. Diffusion MRI has been used to study a variety of different botanical species; for example, maize stems [5,6], barley seeds [7], carrot roots [8], celery stems [9,10] and asparagus stems [11,12].

A recent study has produced diffusion tensor (DT) images (i.e. three-dimensional diffusion images) of the grape berry [13]. While the primary focus of this study was the grape berry vascular system, the orientation of anisotropic diffusion of the berry pericarp was also visible. However, the anisotropy of these observed diffusion patterns did not appear to be related to the known pericarp tissue structure. This is contrary to the commonly shared expectation that anisotropic diffusion should correlate with the orientation of organised tissue structures [14,15].

The objective of the current study was to verify whether diffusion anisotropy features can reflect the relative size, orientation and organisation of the cells that constitute the different tissues of the grape berry pericarp. As previous MRI studies of grape berry tissue structure have focussed on grape berries at single time-points, the second objective of the study was to examine structural changes across grape berry development via diffusion and relaxation MRI. Additional experiments using a postharvest olive as a test sample demonstrated that the diffusion and relaxation MRI techniques are applicable to the study tissue structure of fruits with high oil content.

Since the interpretation of the results of the current investigation require some understanding of relaxation and diffusion MRI, the salient points of both imaging techniques are presented in the following section. More detailed coverage of these topics is available elsewhere [14,16].

## Transverse Relaxation and Diffusion MRI Theory

MRI image contrast depends on the pulse sequence used to weight the measured magnetic resonance (MR) signal. The two forms of MR image contrast used in the current study were the rate of transverse magnetic relaxation and translational self-diffusion.

The transverse relaxation (i.e., spin-spin relaxation) of magnetisation relates to the rate at which the transverse component of the nuclear spin magnetisation vanishes after the system has been excited by the application of electromagnetic radio frequency pulses [17]. The relaxation rate is mainly related to the reorientational motions of the molecule (or that part of the molecule containing the nuclear spin) with slowly moving nuclei (i.e. being in a solid or more viscous environment) relaxing more quickly. Unbound paramagnetic ions (e.g. iron and manganese ions), can also increase the rate at which spins relax. Transverse relaxation weighted images are commonly measured using a Carr-Purcell-Meiboom-Gill (CPMG) MRI pulse sequence [18,19].

Translational self-diffusion, on the other hand, is the random, thermal motion of freely moving molecules such as water. This is distinct from mutual diffusion (i.e. molecular motion due to the presence of a chemical potential). For molecules in bulk solution, translational self-diffusion (henceforth referred to as 'diffusion') can be described using the scalar diffusion coefficient ($D$). For water molecules in bulk solution at standard room temperature, $D \approx 2.29 \times 10^{-9}$ m$^2$ s$^{-1}$[20]. The root mean square displacement (RMSD) of a species diffusing in one dimension is given by

$$\sqrt{2D\Delta} \tag{1}$$

where $\Delta$ is the timescale of the measurement [21]. Water molecules in bulk solution at room temperature diffusing over a period of 50 ms have a RMSD of ~15 μm (i.e. the length scale of eukaryotic cell compartments [22]). The pulsed gradient spin-echo (PGSE) diffusion MRI sequence measures the displacement of the diffusing molecules along the direction of the applied magnetic field gradient used in the sequence [14,23,24]. The intensity of the diffusion weighted MR signal is given by;

$$\ln(S / S_0) = \ln(E) = -\gamma^2\delta^2g^2(\Delta - \delta / 3)D \tag{2}$$

where $S$ is the diffusion weighted MR signal intensity, $S_0$ is the diffusion unweighted MR signal intensity, $E$ is the MR signal attenuation, $\gamma$ is the gyromagnetic ratio of the nuclei of interest, $g$ is the diffusion magnetic gradient strength, $\delta$ is the length of the applied magnetic field gradient pulse, and $\Delta$ is the diffusion period. The diffusion weighting of the pulse sequence can be summarised as a scalar factor, $b$[14,23], where

$$b = \gamma^2\delta^2g^2(\Delta - \delta / 3) \tag{3}$$

When diffusing molecules encounter physical obstructions (e.g. cellular boundaries) their molecular displacement is restricted [14,25]. Consequently, if $D$ is computed from a measured RMSD using Eq. (1) an 'apparent' diffusion coefficient (ADC), which is equal to or smaller than the bulk $D$, will be determined. The ADC of water measured across a span of cells is dependent on various morphological features, such as cell size and orientation, and the permeability of the cellular membrane [26]. As such, the measured ADC will depend upon the direction of measurement. If molecular mobility is equal in all directions, the apparent diffusion is isotropic and a single ADC is sufficient to characterise diffusion. Conversely, if mobility is not equal in all directions due to the arrangement of microscopic tissue structures, the apparent diffusion is anisotropic (Figure 1). The direction of greatest diffusion mobility (i.e. the direction of least diffusive restriction) can be indicated with a diffusion vector. In the case of anisotropic diffusion, the directional dependence of diffusion can be included in Eq. (2) by replacing $D$ with a diffusion tensor, $\mathbf{D}$[24,27]. In order to reconstruct $\mathbf{D}$, PGSE measurements along six or more non-collinear directions are performed. This form of MRI is termed diffusion tensor imaging (DTI).

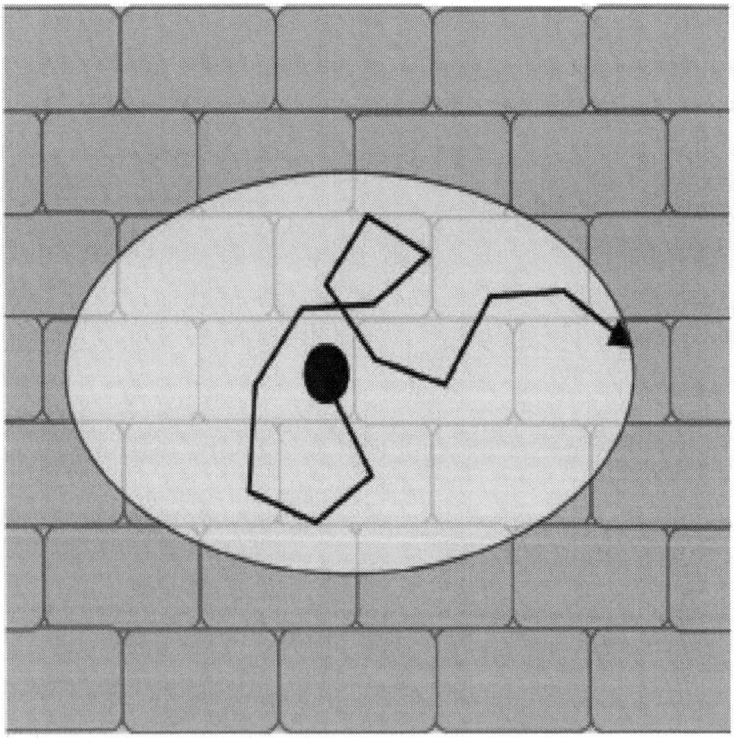

**Figure 1:** Anisotropic water diffusion due to diffusive restriction (two dimensional representation). Here water molecules (•) encounter cellular boundaries as they diffuse. As there are more cellular boundaries on the longitudinal axis than the lateral axis, water displacement along the longitudinal axis is reduced relative to the lateral axis (represented by the yellow ellipse).

# METHODS

## Grape Berry Growing Conditions

*Vitis vinifera* L. cv. Semillon (clone DA16162, own roots) berries were obtained from 18 grapevines grown in a glasshouse at the National Wine & Grape Industry Centre, Wagga Wagga, NSW, Australia. The plants were maintained in 35 L pots containing sandy loam. In late winter each plant was pruned to a pair of two-bud spurs and brought into the glasshouse to commence a fifth season of vegetative growth. Average daily maximum/minimum temperature and relative humidity in the glasshouse across the growing season were 30/17°C and 75/40%, respectively. Each plant was watered beyond field capacity four times per day via a pair of drip emitters. Diluted fertiliser (Megamix Plus, Rutec Pty Ltd, Tamworth, Australia, 20 mL per plant) was applied fortnightly to the soil and wettable sulphur/tribasic copper sulphate was sprayed on the shoots. The shoots were trained vertically and pruned to one inflorescence each approximately six weeks after bud burst. The date of bud burst was assessed on each bud [28] and the date of flowering (approximately 100% capfall) was assessed on each inflorescence.

## MRI Hardware and Software

All $^1$H MRI was performed on a 500 MHz (11.7 T) wide-bore nuclear magnetic resonance spectrometer (AVANCE II; Bruker Biospin Co., Ltd., Germany), equipped with triple axis gradients capable of generating 1.5 T m$^{-1}$ magnetic field gradients. Each imaging experiment employed a 30 mm birdcage radio frequency coil insert. Standard Bruker sequences were used (see Experimental procedures). Data acquisition, post-processing and imaging was controlled from a computer terminal running a Linux operating system, using ParaVision (5.1; Bruker Biospin Co., Ltd.). Image analysis and additional post-processing was performed on a separate computer terminal running Microsoft Windows 7. To analyse the images, the raw MRI data were imported into MATLAB (8.0.0.783, the MathWorks, USA) for analysis using in-house programs.

## Experimental Procedures

Berries were sampled fortnightly for imaging, beginning four weeks after flowering and ending three months later. To minimise developmental variability at each sampling event, grape bunches were assigned to classes according to flowering date (28, 41, 55, 70, 84, 95 and 109 days after flowering (DAF) respectively). Seven bunches were randomly chosen per sampling event. Each sample consisted of the distal portion of the bunch (approximately 10 berries), which were obtained by cutting the rachis while it was momentarily submersed in tap water. The detached, distal portion of the bunch was wrapped in a moist paper towel, placed in a zip-lock bag and shipped overnight to the Biomedical Magnetic Resonance Facility located at the University of Western Sydney, Campbelltown, NSW, Australia.

Three berries were examined per age class (for a total of twenty-one grape berries) to monitor changes in the diffusion pattern across the berry over the course of its development. Berries of an average size (with respect to the bunch) were cut 3 to 4 mm above the pedicel. All MRI protocols were completed within 11 hours of berry detachment. After imaging, the berry was dried at 70°C and then weighed. All grape bunches were stored at 4°C while not in use. Once selected for imaging, a grape berry had approximately 30 minutes to equilibrate to room temperature, maintained at 22 ± 0.1°C, during the spectrometer setup and calibration stages. The interior temperature of the spectrometer was also maintained at 22 ± 0.1°C. The total soluble solids (°Brix) was estimated for each grape berry imaged, based on the average of three refractormeter (PAL-1, Atago Co., Ltd., Tokyo, Japan) readings obtained from berries of the same bunch as the berry imaged.

Each berry was imaged using the MRI sequences described below. Pulse sequence parameters include the slice thickness (THK); the matrix size (MTX), which indicates the number of voxels (i.e. volume elements) which comprise the image; the repetition time (TR) and the echo time (TE), which can be independently adjusted to alter the magnetisation relaxation weighting of the image; number of echoes (if applicable) and the number of times the signal was averaged (NA). For most MRI experiments, two slices of the berry were imaged consecutively; one slice with a transverse orientation (through the centre of the berry, perpendicular to the central vascular bundles) and a second with a longitudinal orientation (through the centre of the berry, parallel to the central vascular bundles). The field of view (FOV) required adjustment for each berry imaged due to the variation in berry size. As a result, the voxels of each image differ in volume (as indicated in the figure captions), although all voxels had a volume less than 0.02 $mm^3$. This is in contrast to clinical MRI scanners which have voxel volumes of 1 – 8 $mm^3$.

A standard multi-shot multi-echo imaging (MSME) sequence (a CPMG based imaging sequence) was employed to produce two-dimensional images weighted by transverse relaxation. The sequence parameters used included THK 1 mm, MTX 256 × 256, a train of 16 echoes spaced 10 ms apart, TR 5 s and NA 2 (total acquisition time ~32 min).

A standard PGSE echo planar DTI sequence [29] was used to produce three-dimensional images weighted by diffusion. The sequence parameters used included THK 1 mm, MTX 128 × 128, $\delta$ 1 ms, $\Delta$ 25 – 50 ms, TE 35 – 65 ms, TR 6500 – 15000 ms and NA 1 (total acquisition time ~4.5 h). The $\Delta$, TE and TR for the DTI sequences were adjusted for each age class of grape berry due to changes in the transverse and longitudinal relaxation properties of the berries across their development (Table 1). The TE was set in order to maximise the length of $\Delta$, and minimise the loss of MR signal due to transverse relaxation. The TR was adjusted to be five times the length of the maximum longitudinal relaxation time ($5 \times T_1$) of each grape berry in order to minimise unintended longitudinal relaxation weighting. Two diffusion measurements ($b$ weighting of 250 and 500 s $mm^{-2}$) were performed along forty-two directions (the vertices of a pentakis icosidodecahedron). One$S_0$ image (which had a negligible diffusion weighting) was also acquired in order to normalise the diffusion weighted images.

**Table 1:** Summary of the **Δ**, TE and TR values used for the DTI sequences

| Berry age (DAF) | Δ (ms) | TE (ms) | TR (ms) |
|---|---|---|---|
| 28 | 25 | 40 | 12500 |
| 41 | 50 | 65 | 15000 |
| 55 | 50 | 60 | 15000 |
| 70 | 50 | 60 | 10000 |
| 85 | 25 | 35 | 8000 |
| 95 | 25 | 35 | 7000 |
| 109 | 25 | 35 | 7000 |

These values were adjusted for each age class of grape berry due to changes in the transverse and longitudinal relaxation properties of the berries across their development.

For the analysis of the DT and transverse relaxation images, each grape berry was divided into four pericarp tissue groups (Figure 2); the exocarp, outer mesocarp, inner mesocarp and septum[30,31]. The exocarp was defined as the outer epidermis and outer hypodermis tissue of the grape berry. The outer mesocarp was considered to be the tissue between the outer hypodermis and the tissue exterior to the peripheral vascular bundles, while the inner mesocarp was considered to be the tissue inwards from the peripheral vascular bundles. The septum was defined as the irregular tissue found at the centre of the grape berry adjacent to the berry seed(s) and locule(s). The endocarp was not considered in the analysis because it was beyond the resolution of the relaxation images. The seed interior was considered separately from the listed pericarp tissues. Each region in the images was analysed independently from all other regions by the application of image masks. These image masks were based on the tissue groups visible in the transverse relaxation images (in the manner explained previously).

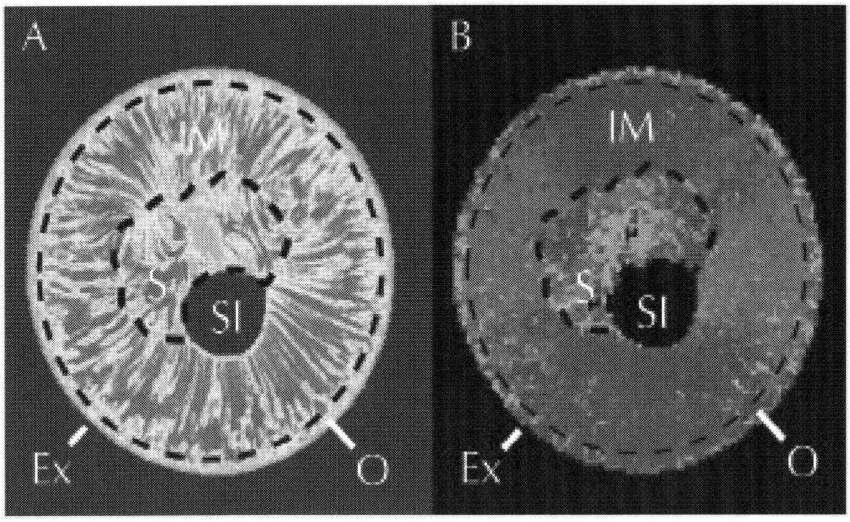


**Figure 2:** Tissue regions of the grape berry (transverse plane). Here the five tissue regions of the grape berry are provided with reference to a transverse relaxation image (A) and a diffusion tensor image (B). Ex: exocarp, OM: outer mesocarp, IM: inner mesocarp, S: septum, SI: seed interior. The outer, black dashed curve indicates the border between the outer mesocarp and the inner mesocarp while the inner, black dashed curve indicates the border between the inner mesocarp and the septum.

For the analysis of the relaxation images, a mean $T_2$ (i.e., spin-spin relaxation time) was determined for each of the pericarp tissues, from each grape berry in the same age group. A standard error was also calculated for each tissue based on the mean $T_2$ values across each respective age group. The acquired DT image data, on the other hand, was used to create three images of each grape berry; a diffusivity map (a map of the average ADC for each voxel), a diffusion vector field map and a diffusion colour map (both of which indicate the direction of least restricted diffusion). The DT colour maps used a symmetrical additive red/blue/green colour scheme to represent that the direction of least restricted diffusion in each voxel was lateral/longitudinal/perpendicular with respect to the FOV. Mean diffusivity values and accompanying standard errors were calculated from the diffusivity maps on a per tissue basis in the same manner as the transverse relaxation maps.

A high spatial resolution DT image of a grape berry 55 DAF was used to confirm whether the radial striation patterns visible in the relaxation images were linked to variations in cellular size across the inner mesocarp tissue. This was done by using the $S_0$ image (which has a small $T_2$ weighting, and had visible striation patterns in the inner mesocarp region) to create an MR signal intensity threshold mask. The threshold value was qualitatively chosen in order to divide the inner mesocarp into radial bands of voxels with 'relatively high' and 'relatively low' signal intensity. The image mask was then applied to the DT image data to create two populations of voxels. The three eigenvalues of the DT (i.e. three orthogonal ADCs, the largest of which corresponds to the direction of least diffusion restriction) for each voxel were averaged over the two populations. This resulted in two groups of primary, secondary and tertiary eigenvalues. The mean primary eigenvalue was proportional to the average length of the cells in the voxel, while the mean secondary and tertiary eigenvalues corresponded to either the average width or depth of the cells. The Tukey-Kramer test ($P$ =0.05) was then performed to determine whether differences between corresponding mean eigenvalues were statistically significant. A statistically significant difference between the values would indicate that there is a significant difference in average cell size across the bands of the inner mesocarp striation patterns.

A DT and transverse relaxation image of a postharvest olive (*Olea europaea* L., cv. Correggiolla) was also acquired in order to determine the applicability of diffusion MRI to fruits with high oil contents. The olive was stored and imaged under the same conditions as the grape berries. The transverse relaxation images were obtained using the MSME pulse sequence and parameters outlined above. The DT images, on the other hand, were acquired using a pulsed gradient stimulated echo (PGSTE) pulse sequence. The sequence parameters used included THK 1 mm, MTX 64 × 64, $\delta$ 6 ms, $\Delta$ 80 ms, TE 20 ms, TR 3500 ms and NA 4 (total acquisition time ~10 h). Two diffusion measurements ($b$ weighting of 80000 and 160000 s mm$^{-2}$) were performed along twenty directions (the vertices of a dodecahedron). One $S_0$ image was also acquired in order to normalise the diffusion weighted images. For the analysis of the transverse relaxation and DT images of the olive, the pericarp was treated as a single tissue. This was due to the larger voxel size of the images and lack of distinctive tissue groups.

# RESULTS

The grape berries increased in size and weight and then declined as the berries progressed through the ripening and senescence phases (Figure 3). The post-harvest olive had a fresh weight of 3.30 g and dry weight of 1.63 g. The concentration of soluble solids in the grape berries increased sigmoidally with respect to time (adjusted $R^2 = 0.99$). Véraison occurred at approximately 60 – 65 DAF, and full ripeness was placed at 95 DAF (based on the mean concentration of soluble solids of the sampled berries which plateaued at 26 °Brix).

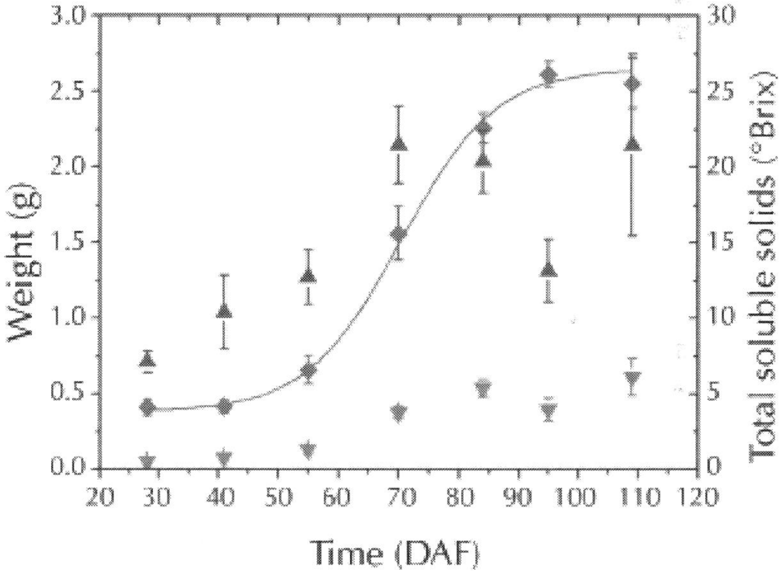


**Figure 3:** The physical characteristics of the grape berries. The concentrations of soluble solids of the grape berries (Green ♦), as well as the fresh weight (Blue ▲) and dry weights (Red ▼) of the berries, are presented with respect to the number of days after flowering. A sigmoidal function (solid green curve) of the form $a_1 + (a_2 - a_1)/(1 + \exp(-(DAF - x_0)/w))$ was fitted to the soluble solids values by nonlinear regression (adjusted $R^2 = 0.99$), where $a_1 = 26.1$ (the approximate maximum soluble solids value), $a_2 = 3.9$ (the approximate minimum soluble solids value), $x_0 = 69.7$ (the inflection point) and $w = 7.4$ (the change in DAF which yielded the greatest change in the soluble solids value). The error bars are given by the standard deviation of soluble solids values at each time point.

In the transverse relaxation images, the exocarp was difficult to distinguish from the outer mesocarp (Figures 4 and 5). The exocarp was consistently associated with the shortest mean $T_2$at all sampled stages of grape berry development, relative to the other tissues examined (Figure 6). The diffusion vectors of the exocarp were prominently aligned tangential to the berry surface (Figures 7, 8 and 9). This pattern was consistent for the exocarp of all grape berries imaged, regardless of berry age. The exocarp was also found to have short mean diffusivity values relative to the other tissues (Figure 10).

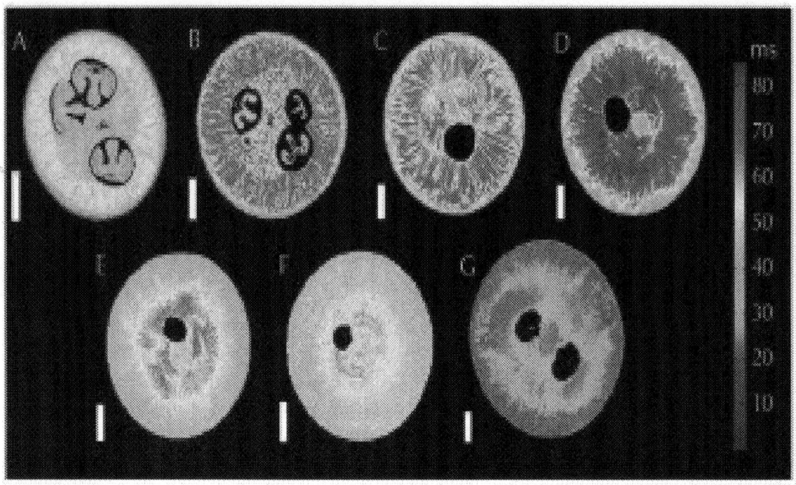


**Figure 4:** Transverse relaxation images of grape berries at seven different stages of berry development (transverse plane). The images include three pre-véraison grapes, at 28 DAF (A, voxel size 59 × 59 × 1000 μm), 41 DAF (B, voxel size 78 × 78 × 1000 μm) and 55 DAF (C, voxel size 78 × 78 × 1000 μm), a grape undergoing véraison at 70 DAF (D, voxel size 82 × 82 × 1000 μm), a ripening grape at 85 DAF (E, voxel size 74 × 74 × 1000 μm), a grape which is at oenological maturity at 95 DAF (F, voxel size 63 × 63× 1000 μm) and a post-maturity berry at 109 DAF (G, voxel size 86 × 86 × 1000 μm). The transverse relaxation values are indicated by the colour bar to the right of the figure. Scale bar: 3 mm.

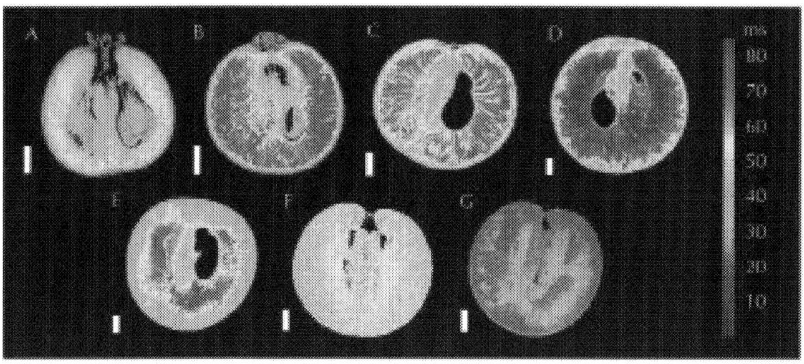


**Figure 5:** Transverse relaxation images of grape berries at seven different stages of berry development (longitudinal plane). The images include three pre-véraison grapes, at 28 DAF (A, voxel size 59 × 59 × 1000 µm), 41 DAF (B, voxel size 78 × 78 × 1000 µm) and 55 DAF (C, voxel size 78 × 78 × 1000 µm), a grape undergoing véraison at 70 DAF (D, voxel size 133 × 133 × 1000 µm), a ripening grape at 85 DAF (E, voxel size 82 × 82 × 1000 µm), a grape which is at oenological maturity at 95 DAF (F, voxel size 63 × 63 × 1000 µm) and a post-maturity berry at 109 DAF (G, voxel size 86 × 86 × 1000 µm). The transverse relaxation values are indicated by the colour bar to the right of the figure. Scale bar: 3 mm.

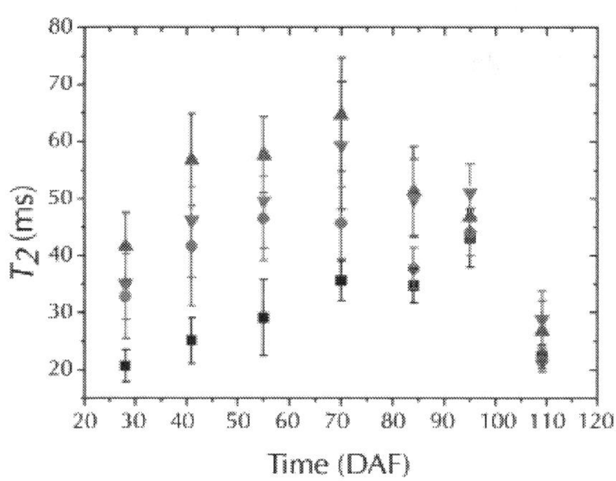


**Figure 6:** Transverse (T2) relaxation of the major tissue groups of the grape berry with respect to time (transverse plane). The transverse relaxation values for the exocarp (■), outer mesocarp (Red ●), inner mesocarp (Blue ▲) and septum (Pink ▼). The error bars are given by the standard deviation of the transverse relaxation values at each time point.

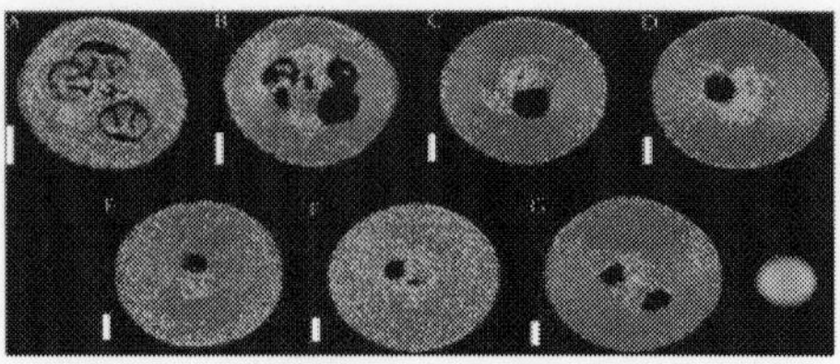

**Figure 7:** DT images of grape berries at seven different stages of berry development (transverse plane). The images include three pre-véraison grapes, at 28 DAF (A, voxel size 117 × 117 × 1000 μm), 41 DAF (B, voxel size 156 × 156 × 1000 μm) and 55 DAF (C, voxel size 156 × 156 × 1000 μm), a grape undergoing véraison at 70 DAF (D, voxel size 164 × 164 × 1000 μm), a ripening grape at 85 DAF (E, voxel size 148 × 148 × 1000 μm), a grape at oenological maturity at 95 DAF (F, voxel size 125 × 125 × 1000 μm) and a post-maturity berry, at 109 DAF (G, voxel size 171 × 171 × 1000 μm). The colours in the figure indicate the direction of least restricted diffusion, as indicated by the image at the bottom right of the figure. Scale bar: 3 mm.

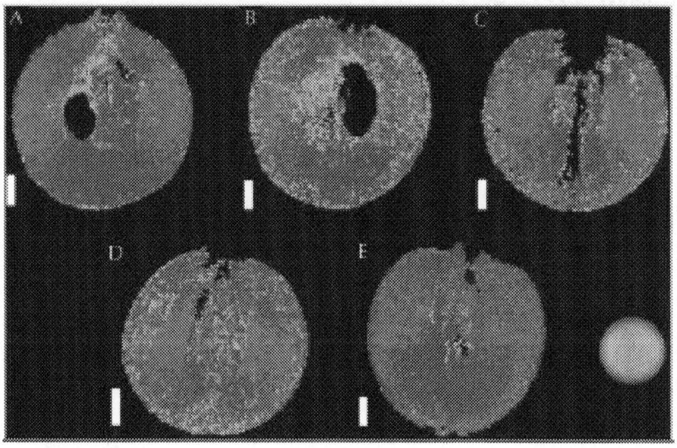

**Figure 8:** DT images of grape berries at five different stages of berry development (longitudinal plane). The images include a pre-véraison grape at 55 DAF (A, voxel size 156 × 156 × 1000 μm), a grape undergoing véraison at 70 DAF (B, voxel size 164 × 164 × 1000 μm), a ripening grape at 85 DAF (C, voxel size 172 × 172 × 1000 μm), a grape which is at oenological maturity at 95 DAF (D, voxel size 125 × 125 × 1000 μm) and a post-maturity berry at 109 DAF (E, voxel size 172 × 172 × 1000 μm). No images are available for 28 and 41 DAF. The colours in the figure indicate the direction of least restricted diffusion, as indicated by the image in the bottom right side of the figure. Images are not available for 28 and 41 DAF. Scale bar: 3 mm.

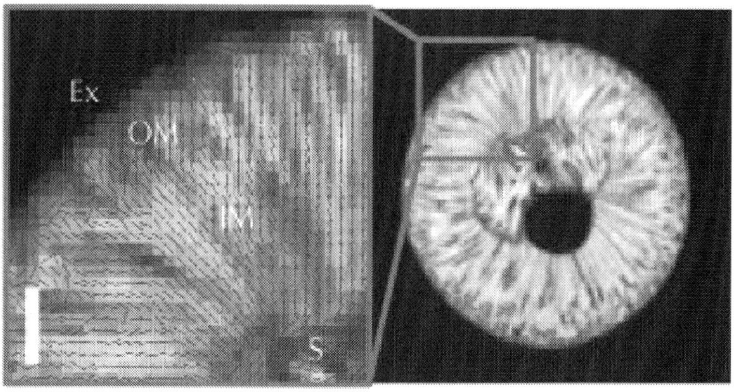


**Figure 9:** Diffusion vector field map overlaying the S0 image of a grape berry 41 DAF (transverse plane). The diffusion vectors (blue arrows) indicate the direction of least restricted diffusion in each voxel. Voxel size: 156 × 156 × 1000 μm, bar length: 1000 μm.

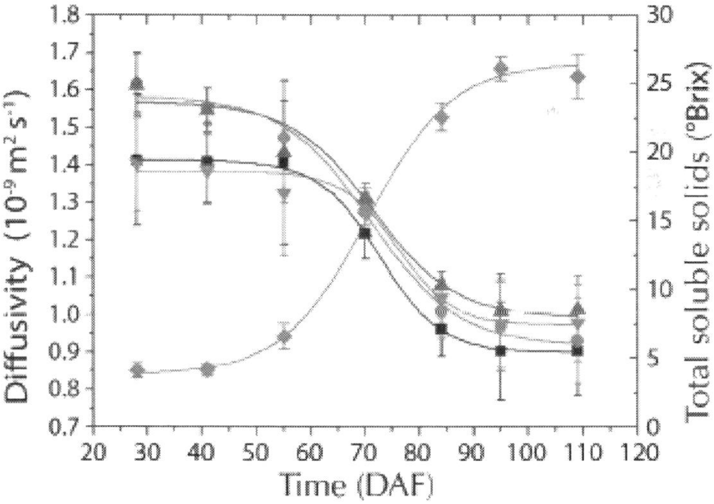


**Figure 10:** Mean ADC of the major tissue groups of the grape berry with respect to time and total soluble solids. The mean ADC for the exocarp (■), outer mesocarp (Red ●), inner mesocarp (Blue ▲) and septum (Pink ▼) decrease sigmoidally (adjusted $R^2 = 0.99$) with respect to sigmoidally increasing (adjusted $R^2 = 0.99$) dissolved solids content (Green ♦). The error bars are given by the standard deviation of the mean ADC at each time point.

The transverse relaxation images of the outer mesocarp generally lacked consistent or well defined features (Figures 4 and 5). The mean $T_2$ of the outer-mesocarp tended to be longer than those of the exocarp, but shorter than other tissues (Figure 6). The diffusion vectors had a rotational dependency (Figures 7, 8 and 9) and were aligned so as to radiate from the centre of the fruit. In pre-véraison berries, the diffusivity values of the outer mesocarp were amongst the highest observed. For post-véraison grape berries, however, the mean diffusivity values declined further than other tissues (Figure 10).

A notable feature in the transverse relaxation images of the inner mesocarp was the clear radial striation patterns with distinctly different (Tukey-Kramer test, $P$ =0.05) $T_2$ values.

These patterns were observed in both transverse and longitudinal image orientations for the inner mesocarp (Figures 4 and 5) of grape berries between 28 DAF and 109 DAF. The striation pattern radiated throughout the entire sub-tissue, starting close to the septum and vascular systems at the centre of the berry (i.e., at the ovular and axial vascular network), and terminating at the interface between the inner and outer mesocarp. In addition, upon analysing the high resolution DT image of a grape berry 55 DAF (Figure 11), a statistically significant difference (Tukey-Kramer test, P =0.05) was noted between the mean secondary and tertiary eigenvalues across the striation bands (Table 2). Prior to 95 DAF, the mean $T_2$ of the inner mesocarp was consistently higher than other tissues (Figure 6). After the concentration of soluble solids plateaued, the mean $T_2$ of this tissue declined, accompanied by a partial loss of its radial striation pattern.

The diffusion vectors of the inner mesocarp had a rotational dependency similar to that of the outer mesocarp. Furthermore, the diffusion vectors were predominantly parallel to the radial striation bands (Figures 7, 8 and 9), except for grape berries 28 DAF or past 95 DAF (Figure 12). After véraison, the inner mesocarp was also consistently associated with the largest mean diffusivity values, relative to the other tissues examined (Figure 10).

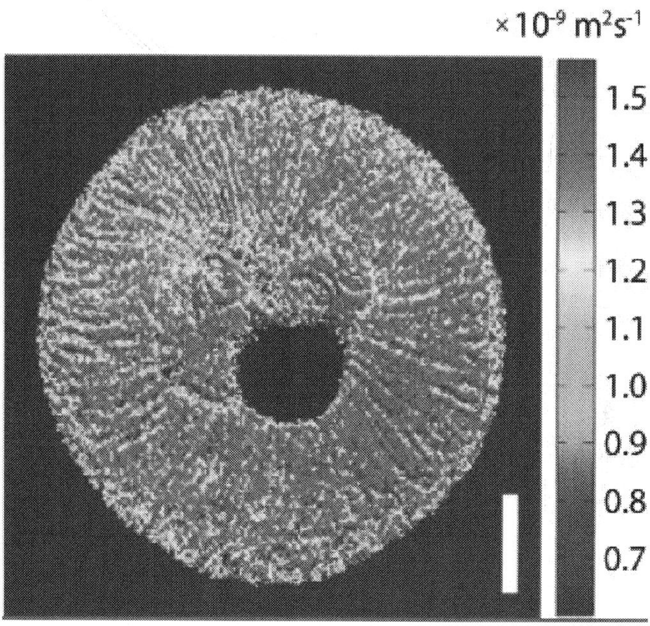


**Figure 11:** Mean ADC map of a grape berry 55 DAF (transverse plane). Voxel size 78 × 78 × 1000 μm, bar length: 3000 μm.

**Table 2:** Comparing the mean eigenvalues of diffusion tensors associated with different relaxation striation bands

| | **Mean primary eigenvalue ($\times 10^{-9}\ m^2\ s^{-1}$)** | **Mean secondary eigenvalue ($\times 10^{-9}\ m^2\ s^{-1}$)** | |
|---|---|---|---|
| **Striation bands with high signal intensity** | 1.53* | 1.44 | 1.36 |
| **Striation bands with low signal intensity** | 1.54* | 1.41 | 1.30 |

The standard error was not included as it was insignificant (<1% of the values listed). * : values between which there was no statistically significant difference (Tukey-Kramer test, $P$ =0.05).

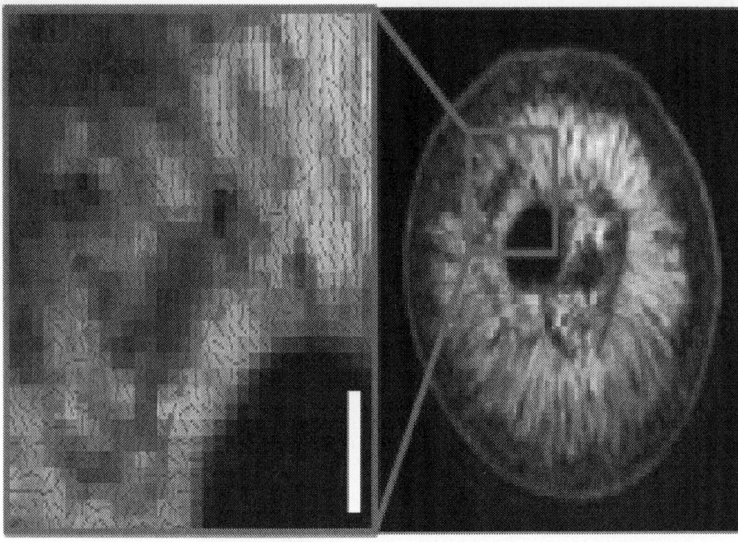

**Figure 12:** Diffusion vector field map overlaying the S0 image of a grape berry 109 DAF (transverse plane). The orientation of diffusion vectors are indicated by the blue arrows. There was a loss of diffusion-weighted signal in the region denoted by the red dashed line. Voxel size: 133 × 133 × 1000 μm, bar length: 1000 μm.

The septum was readily distinguishable from the surrounding mesocarp tissue in the transverse relaxation images. This was due to the presence of curved striation patterns which were perpendicular to the striation patterns of the inner mesocarp (Figures 4 and 5). These striation patterns were curled about the central vascular bundles and extended through the septum to the seeds and locules. Prior to 95 DAF, the mean $T_2$ of the septum was consistently higher than the exocarp, but lower than the inner mesocarp. After 95 DAF, the septum had the highest mean $T_2$(Figure 6). Unlike the inner mesocarp, the diffusion vectors of the septum were not consistently parallel to the striation patterns of the tissue (Figures 7, 8 and 9). The anisotropic diffusion pattern of the septum differed on a per berry basis. The septal-mesocarp boundary could be discerned by abrupt changes in the local orientation of diffusion vectors. The mean diffusivity value of the septum was the lowest of all the examined tissues pre-véraison, but had a mean diffusivity greater than that of the exocarp and outer mesocarp post-véraison (Figure 10).

In the grape seed interior, the nucellus (located towards the bulbous distal end of the seed) was identified in grape berries aged 28 and 41 DAF. It could be readily identified in berries of this age, as it occupies a sizeable portion of the seed and has a stylised '3' shape when imaged through the transverse plane [4]. The nucellus demonstrated a highly characteristic anisotropic diffusion pattern which was rotationally dependent, similar to the anisotropic diffusion pattern of the inner mesocarp tissue (Figures 7, 8 and 13). After 41 DAF, the MR signal from this region decayed too rapidly to be visible in the DT and transverse relaxation images.

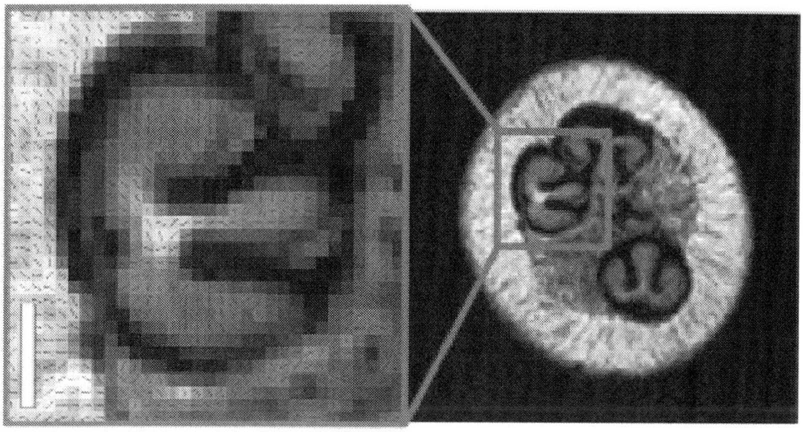

**Figure 13:** Diffusion vector field map overlaying the S0 image of a grape berry seed interior 28 DAF (transverse plane). The diffusion vectors (blue arrows) indicate the direction of least restricted diffusion. Voxel size: 117 × 117 × 1000 μm, bar length: 1000 μm.

The transverse relaxation images of the postharvest olive pericarp lacked well-defined macroscopic features, such as the radial striation patterns noted in the mesocarp of the grape berries. However, the anisotropic diffusion patterns (Figure 14A) and diffusion vectors (Figure 14B) associated with the olive pericarp exhibited a rotational dependency and were aligned so as to radiate from the centre of the fruit. The mean $T_2$ of the olive pericarp was 45 ± 3 ms while the mean diffusivity value of the olive pericarp was $6.07 \pm 0.46 \times 10^{-12}$ $m^2$ $s^{-1}$.

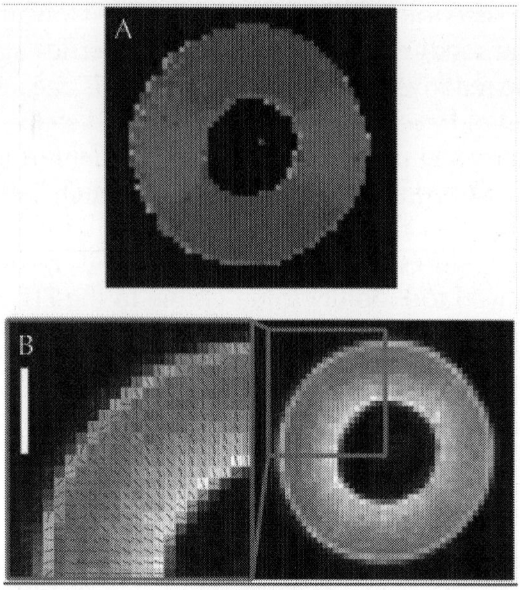


**Figure 14:** DT and transverse relaxation images of a postharvest olive (transverse plane). The DT image (A, voxel size 390 × 390 × 1000 μm) and diffusion vector map (B, voxel size 390 × 390 × 1000 μm) of an olive pericarp. Bar length 1000 μm.

# DISCUSSION

The anisotropic diffusion patterns observed in the grape berry pericarp were due to the restricting effects of cell membranes on diffusion. For example, in the mesocarp of grape berries 28 DAF, the diffusion anisotropy exhibited low coherence. This was because the parenchyma cells of the mesocarp were not fully elongated [32], as demonstrated by confocal microscopy (Figure 15). Between 41 DAF and 95 DAF, however, the anisotropic diffusion pattern was clearly radially dependent, thus reflecting the radial orientations of the elongated inner mesocarp cells [3,30,32], as shown by confocal microscopy (Figure 16). The orientations of the diffusion vectors in the inner mesocarp were also preferentially parallel to the radial striation patterns visible in the transverse relaxation images, a relationship that will be discussed further below.

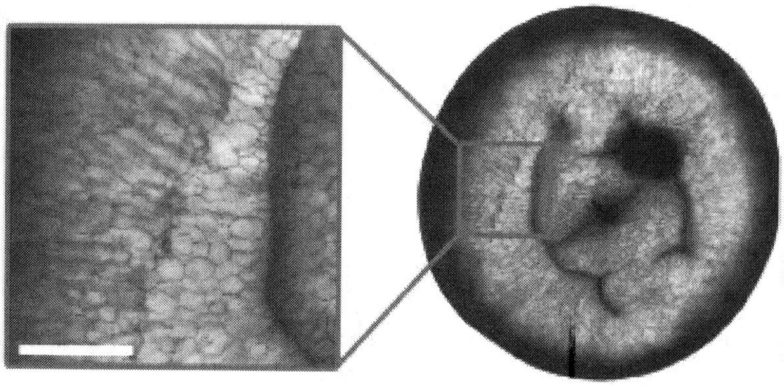

**Figure 15:** Confocal micrograph of the pericarp of a grape berry prior to véraison 41 DAF (transverse plane). The image was acquired using a confocal microscope (LSM5 Pascal; Zeiss, Germany) which employed a 488 nm Argon laser and a 10 × objective Plan-Apochromatic lens. Bar length 1000 μm.

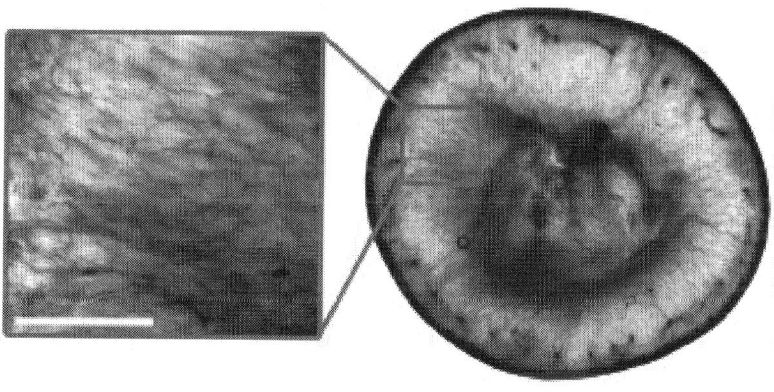

**Figure 16:** Confocal micrograph of the pericarp of a grape berry undergoing véraison 55 DAF (transverse plane). The image was acquired using a confocal microscope (LSM5 Pascal; Zeiss, Germany) which employed a 488 nm Argon laser and a 10 × objective Plan-Apochromatic lens. Bar length 1000 μm.

Upon approaching full ripeness, regions of the berry inner mesocarp exhibited a loss of coherence in orientation of diffusion anisotropy (Figure 12). This was accompanied by a decline in $T_2$ values and partial loss of the radial striation pattern in the transverse relaxation images of these same regions. The decline of $T_2$ values would begin at the

than the cells that make up the mesocarp, thus a short Δ relative to that used in the current study remains appropriate for probing the cellular membranes of the xylem and phloem. However, in regions where the average cell size is much larger, (e.g. the inner mesocarp) the diffusion anisotropy patterns are more likely to be observed due to interactions with cellular components rather than the cellular membrane. As a result, the observations from both the current study and that of Gruwel et al. could be considered accurate; however, they are observations of the effects of different restricting structures (i.e., on a shorter length scale).

The second difference between the present study and that of Gruwel et al. was the diffusion gradient schemes employed. Gruwel et al. performed diffusion measurements along six unique diffusion gradient vectors. While this is the minimum required for DTI, it is possible that gradient cross-terms (from the imaging and diffusion gradients) have affected the diffusion weighting of the images. The current study used forty-two different diffusion gradient vectors (i.e., twenty-one unique directions and their corresponding opposite). The inclusion of gradient vectors and their corresponding opposite gradient vectors reduces the effects of potential gradient cross-terms [27,40]. Using a large number of unique diffusion gradient vectors also reduces potential directional bias when reconstructing D.

## The Applicability of Diffusion MRI to Plant Tissues Differing in Oil and Water Content

The question remains as to the applicability of diffusion MRI in the examination of other plant species and organs. This will chiefly depend upon the amount of unbound water the organ contains. Tissues with low content, such as the solid endosperm of seeds, are difficult to examine. In addition, tissues with smaller cells are easier to examine than tissues with larger cells. However, providing that Δ is long enough to ensure that a large population of water molecules are able to interact with the surface of the cellular boundaries (e.g. $\Delta \sim d^2/D$ where $d$ is the distance between cellular boundaries), diffusion anisotropy will be evident.

The applicability of diffusion MRI to the study of plant species with high oil contents is of great interest; oil is hydrophobic and thus diffusing water molecules are likely to treat any droplets of oil encountered in cell cytoplasm [41] as diffusion restricting obstacles. Depending on the concentration of oil in the cytoplasm, water diffusivity could be noticeably affected. Olives are an ideal candidate for such a study as they are comparable in size to grape berries and also have a similar internal tissue structure and cellular dimensions. These features would aid comparisons to the results of the current study. To determine the practicality of such an investigation, a postharvest olive was imaged using both DT and transverse relaxation MRI.

Due a noticeable loss of radio-frequency magnetic pulse power, fast diffusion MRI sequences could not be used, increasing the length of the total image acquisition. This was a consequence of the high salt content of the fruit. The low mean diffusivity value of olive pericarp also necessitated the use of very high diffusion magnetic gradients to perform the diffusion measurements. A lengthy diffusion period was also required to ensure that the slowly diffusing molecules had sufficient time to probe the restricting tissue microstructure. To prevent prohibitive MR signal loss arising from transverse relaxation, a PGSTE MRI sequence was employed. While the two pulse sequences measure diffusion in the same way, the PGSTE sequence halves the intensity of the acquired MR signal. The number of image averages and the size of the image voxel were thus increased to compensate for this loss of signal.

The resultant diffusion MRI images had voxel volume less than 1 $mm^3$ and a good signal-to-noise ratio. The radial anisotropic diffusion pattern observed in the olive pericarp was similar to the grape berries close to or past véraison, indicating that the cells of the mesocarp are likewise radially elongated. This result agrees with what is known of the olive mesocarp structure [42]. The mean $T_2$ value of the olive was comparable to those of the grape berry pericarp tissues, as was expected for fruits with similar cellular dimensions. The large difference between the mean ADC of the olive pericarp and the grape berry pericarp tissues is likely due to the presence of oil in the cell cytoplasm. However a more in-depth investigation is required to speculate further.

# CONCLUSION

The results of the current study have demonstrated that diffusion anisotropy patterns are correlated with the microstructure of the major pericarp tissues of cv. Semillon grape berries, including the exocarp, outer and inner mesocarp, and seed interior. Changes in grape berry tissue structure during berry development, which were observable in the DT and transverse relaxation images, have also been described. These changes in tissue structure included the transition from non-elongated to radially elongated mesocarp cells between 28 and 41 DAF, which was associated with an increase in diffusion vector coherence. Conversely, from the mid-ripe phase and beyond, the loss of diffusion vector coherence was possibly the result of the widespread cell death in the mesocarp that often accompanies this stage of development in wine grapes. Rotationally dependent anisotropic diffusion patterns were also noted in the seed integument of young grape berries. These diffusion patterns could only be observed early in the development of the grape berry, due to the dehydration of the seed integument and the replacement of the nucellus by the liquid endosperm.

The results of the current study agree that the inner mesocarp striation patterns previously noted in the transverse relaxation images of previous studies arise due to variations in cell width across the striation bands. These bands were evident in the images of the grape berries at 28 DAF and largely persisted through all subsequent phases of berry development.

## ACKNOWLEDGEMENTS

The research is supported by a NSW BioFirst Award from the NSW Ministry for Science & Medical Research (WSP), as well as the grape growers and winemakers of Australia through their investment body, the Australian Grape and Wine Authority, with matching funds from the Federal Government (SYR). RJD gratefully acknowledges the support of the Professor Kai Yip Cho Memorial Scholarship. The authors acknowledge the facilities, and the scientific and technical assistance of the National Imaging Facility, University of Western Sydney Node.

## REFERENCES

1. Bernstein M, King K, Zhou X, Fong W: Handbook of MRI pulse sequences. *Med Phys* 2005, 32:1452.

2. Lauterbur P: Image formation by induced local interactions: examples employing nuclear magnetic resonance. *Nature* 1973, 242:190-191.
3. Glidewell SM, Williamson B, Goodman BA, Chudek JA, Hunter G: An NMR microscopic study of grape (Vitis vinifera L.). *Protoplasma* 1997, 198:27-35.
4. Pope J, Jonas D, Walker R: Applications of NMR micro-imaging to the study of water, lipid, and carbohydrate distribution in grape berries. *Protoplasma* 1993, 173:177-186.
5. Van Der Weerd L, Claessens MMAE, Efdé C, Van As H: Nuclear magnetic resonance imaging of membrane permeability changes in plants during osmotic stress. *Plant Cell Environ* 2002, 25:1539-1549.
6. Kuchenbrod E, Haase A, Benkert R, Schneider H, Zimmermann U: Quantitative NMR microscopy on intact plants. *Magn Reson Imaging* 1995, 13:447-455.
7. Ishida N, Ogawa H, Kano H: Diffusion of cell-associated water in ripening barley seeds. *Magn Reson Imaging* 1995, 13:745-751.
8. Sibgatullin T, Vergeldt F, Gerkema E, Van As H: Quantitative permeability imaging of plant tissues. *Eur Biophys J* 2009, 39:699-710.
9. Trudeau JD, Dixon WT, Hawkins J: The effect of inhomogeneous sample susceptibility on measured diffusion anisotropy using NMR imaging. *J Magn Reson B* 1995, 108:22-30.
10. Sotak CH: Nuclear magnetic resonance (NMR) measurement of the apparent diffusion coefficient (ADC) of tissue water and its relationship to cell volume changes in pathological states. *Neurochem Int* 2004, 45:569-582.
11. Boujraf S, Luypaert R, Eisendrath H, Osteaux M: Echo planar magnetic resonance imaging of anisotropic diffusion in asparagus stems. *MAGMA* 2001, 13:82-90.
12. Lätt J, Nilsson M, Rydhög A, Wirestam R, Ståhlberg F, Brockstedt S: Effects of restricted diffusion in a biological phantom: a q-space diffusion MRI study of asparagus stems at a 3 T clinical scanner. *MAGMA* 2007, 20:213-222.

13. Gruwel MLH, Latta P, Sboto-Frankenstein U, Gervai P: Visualization of water transport pathways in plants using diffusion tensor imaging. *Prog Electromagn Res C* 2013, 35:73-82.
14. Price WS: *NMR studies of translational motion*. Cambridge: Cambridge University Press; 2009.
15. Le Bihan D, Mangin J, Poupon C, Clark C, Pappata S, Molko N, Chabriat H: Diffusion tensor imaging: concepts and applications. *J Magn Reson Imaging* 2001, 13:534-546.
16. Levitt M: *Spin dynamics: Basics of nuclear magnetic resonance*. 2nd edition. New York: Wiley; 2001.
17. Hahn EL: Spin Echoes. *Phys Rev* 1950, 80:580-594.
18. Meiboom S, Gill D: Modified spin-echo method for measuring nuclear relaxation times. *Rev Sci Instrum* 1958, 29:688-691.
19. Carr HY, Purcell EM: Effects of diffusion on free precession in nuclear magnetic resonance experiments. *Phys Rev* 1954, 94:630-638.
20. Weingärtner H: Self diffusion in liquid water. A reassessment. *Z Phys Chem* 1982, 132:129-149.
21. Einstein A: *Investigations on the theory of the Brownian movement*. New York: Dover Publications; 1956.
22. Mano S, Nakamura T, Kondo M, Miwa T, Nishikawa S-i, Mimura T, Nagatani A, Nishimura M:The plant organelles database 3 (PODB3) update 2014: Integrating electron micrographs and new options for plant organelle research. *Plant Cell Physiol* 2014, 55:e1.
23. Stejskal EO, Tanner JE: Spin diffusion measurements: spin echoes in the presence of a time-dependent field gradient. *J Chem Phys* 1965, 42:288-292.
24. Basser PJ, Mattiello J, LeBihan D: Estimation of the effective self-diffusion tensor from the NMR spin echo. *J Magn Reson B* 1994, 103:247-254.
25. Crank J: *The mathematics of diffusion*. U.S.A.: Oxford University Press; 1979.
26. Marrink S-J, Berendsen HJC: Simulation of water transport through a lipid membrane. *J Phys Chem* 1994, 98:4155-4168.

27. Kingsley PB: Introduction to diffusion tensor imaging mathematics: part II. Anisotropy, diffusion-weighting factors, and gradient encoding schemes. *Concepts Magn Reson A* 2006, 28A:123-154.

28. Coombe BG: Growth stages of the grapevine: adoption of a system for identifying grapevine growth stages. *Aust J Grape Wine R* 1995, 1:104-110.

29. Mansfield P: Multi-planar image formation using NMR spin echoes. *J Phys C: Solid State Phys* 1977, 10:L55.

30. Pratt C: Reproductive anatomy in cultivated grapes - A review. *Am J Enol Viticult* 1971, 22:92-109.

31. Hardie WJ, O'Brien TP, Jaudzems VG: Morphology, anatomy and development of the pericarp after anthesis in grape, *Vitis vinifera* L. *Aust J Grape Wine R* 1996, 2:97-142.

32. Harris JMK PE, Possingham JV: Anatomical aspects of grape berry development. *Vitis* 1968, 7:106-119.

33. Edzes HT, van Dusschoten D, Van As H: Quantitative $T_2$ imaging of plant tissues by means of multi-echo MRI microscopy. *Magn Reson Imaging* 1998, 16:185-196.

34. Van As H, Scheenen T, Vergeldt FJ: MRI of intact plants. *Photosynthesis Res* 2009, 102:213-222.

35. Tilbrook J, Tyerman S: Cell death in grape berries: varietal differences linked to xylem pressure and berry weight loss. *Funct Plant Biol* 2008, 35:173.

36. Krasnow M, Matthews M, Shackel K: Evidence for substantial maintenance of membrane integrity and cell viability in normally developing grape (Vitis vinifera L.) berries throughout development. *J Exp Bot* 2008, 59:849-859.

37. Cadot Y, Miñana-Castelló MT, Chevalier M: Anatomical, histological, and histochemical changes in grape seeds from *Vitis vinifera* L. cv Cabernet franc during fruit development. *J Agric Food Chem* 2006, 54:9206-9215.

38. Mullins M, Bouquet A, Williams L: *Biology of the grapevine.* Cambridge: Cambridge University Press; 1992.

39. Ebadi A, Sedgley M, May P, Coombe BG: Seed development and abortion in *Vitis vinifera*L., cv. Chardonnay. *Int J Plant Sci* 1996, 157:703-712.

40. Neeman M, Freyer JP, Sillerud LO: A simple method for obtaining cross-term-free images for diffusion anisotropy studies in NMR microimaging. *Magn Reson Med* 1991, 21:138-143.
41. Rangel B, Platt KA, Thomson WW: Ultrastructural aspects of the cytoplasmic origin and accumulation of oil in olive fruit (*Olea europaea*). *Physiol Plant* 1997, 101:109-114.
42. Hammami SB, Rapoport HF: Quantitative analysis of cell organization in the external region of the olive fruit. *Int J Plant Sci* 2012, 173:993-1004

Chapter 10

# High Throughput Quantitative Phenotyping of Plant Resistance Using Chlorophyll Fluorescence Image Analysis

Céline Rousseau[1,2,3], Etienne Belin[4], Edouard Bove[1,2,3], David Rousseau[4,5], Frédéric Fabre[6],Romain Berruyer[1,2,3], Jacky Guillaumès[1,2,3], Charles Manceau[7], Marie-Agnès Jacques[1,2,3] and Tristan Boureau[1,2,3,8*]

[1]INRA, UMR1345 Institut de Recherche en Horticulture et Semences, Beaucouzé F-49071, France

[2]UMR1345 Institut de Recherche en Horticulture et Semences, SFR4207 QUASAV, PRES L'UNAM, Université d'Angers, Angers F-49045, France

[3]AgroCampus-Ouest, UMR1345 Institut de Recherche en Horticulture et Semences, Angers, F-49045, France

[4]Université d'Angers, Laboratoire d'Ingénierie des Systèmes Automatisés (LISA), Angers, F- 49000, France

[5]Present address: CREATIS; CNRS UMR5220; INSERM U630, Université de Lyon, Villeurbanne, F-69621, France

[6]INRA, UR0407 Pathologie Végétale, Montfavet, F-84140, France
[7]ANSES, Direction de la Santé des Végétaux, Angers, France
[8]Université d'ANgers, UMR1345 Institut de Recherche en Horticulture et Semences, Beaucouzé, F-49071, France

# ABSTRACT

## Background

In order to select for quantitative plant resistance to pathogens, high throughput approaches that can precisely quantify disease severity are needed. Automation and use of calibrated image analysis should provide more accurate, objective and faster analyses than visual assessments. In contrast to conventional visible imaging, chlorophyll fluorescence imaging is not sensitive to environmental light variations and provides single-channel images prone to a segmentation analysis by simple thresholding approaches. Among the various parameters used in chlorophyll fluorescence imaging, the maximum quantum yield of photosystem II photochemistry ($F_v/F_m$) is well adapted to phenotyping disease severity. $F_v/F_m$ is an indicator of plant stress that displays a robust contrast between infected and healthy tissues. In the present paper, we aimed at the segmentation of $F_v/F_m$ images to quantify disease severity.

## Results

Based on the $F_v/F_m$ values of each pixel of the image, a thresholding approach was developed to delimit diseased areas. A first step consisted in setting up thresholds to reproduce visual observations by trained raters of symptoms caused by *Xanthomonas fuscans* subsp. *fuscans* (*Xff*) CFBP4834-R on*Phaseolus vulgaris* cv. Flavert. In order to develop a thresholding approach valuable on any cultivars or species, a second step was based on modeling pixel-wise $F_v/F_m$-distributions as mixtures of Gaussian distributions. Such a modeling may discriminate various stages of the symptom development but over-weights artifacts that can occur on mock-inoculated samples. Therefore, we developed a thresholding approach based on the probability of misclassification of a healthy pixel. Then, a clustering step is performed on the diseased areas to discriminate between various stages of alteration of plant tissues. Notably, the use of chlorophyll fluorescence imaging could detect pre-symptomatic area. The interest of this image analysis procedure for assessing the levels of quantitative resistance is illustrated with the quantitation of disease severity on five commercial varieties of bean inoculated with *Xff* CFBP4834-R.

## Conclusions

In this paper, we describe an image analysis procedure for quantifying the leaf area impacted by the pathogen. In a perspective of high throughput phenotyping, the procedure was automated with the software R downloadable at http://www.r-project.org/.

# BACKGROUND

Quantitative phenotyping is important in plant breeding to screen accessions for resistance to pathogens. Indeed, plant resistance to pathogens may either be qualitative or quantitative. Qualitative resistance is due to the presence of single major resistance genes that confer total resistance to pathogens carrying the cognate avirulence genes. However, these monogenic total resistances are often rapidly bypassed. In contrast, resistance conferred by multiple loci exhibit quantitative phenotype and is thought to contribute to durability [1-3]. Thus quantitative phenotyping methods are necessary to ensure a good evaluation of the disease severity and to make appropriate decisions in gauging cultivar resistance in plant breeding.

Visual assessments have often been used to quantify disease severity. They require rating scales to be finely tuned, and raters to be trained, which increases considerably the cost and the time devoted to phenotyping projects. Moreover, these methods highly depend on the subjectivity of the raters and thus often lack accuracy, reproducibility and traceability [4-7].

In contrast, automated image analysis-based phenotyping provides a powerful alternative to visual assessments. Indeed, automation eventually provides a calibrated image analysis, thereby eliminating any subjectivity of the raters and ensuring reproducibility [6]. Furthermore, automation allows high throughput phenotyping. Calibrated protocols and data storage may provide useful tools for traceability or inter-laboratory comparison of the results.

Image segmentation aims at partitioning the digital image into multiple sets of pixels to select the areas of interest. In plant pathology, imaging of the symptoms has been mainly achieved using conventional color imaging. Segmentation algorithms aimed at the automated quantification of the total amount of symptoms on leaves from conventional color images [8-12]. Some of them aim at the quantification of the chlorosis or the necrosis using the differential colors of diseased leaf tissues [10,12]. Indeed, yellow and brown coloration respectively characterize chlorotic and necrotic tissues. Such automated segmentation procedures rely on color-based thresholds to discriminate diseased from healthy tissues on leaves. However, heterogeneity during exposition may alter the contrast of visible images and interfere with any color-based thresholding. Batch segmentation of visible images using color-based thresholds may thus generate numerous artifacts if light conditions during exposure are not tightly controlled. Scanner imaging of detached leaves or adjusting the color balance using a color checker may help standardizing contrasts prior to the segmentation procedures [9]. As conventional color images are typically multichannel images (for instance RGB images are composed by red, green and blue channels), they need sophisticated image analysis methods [13]. As segmentation by simple thresholding can only be applied on single channel images, conventional color images should be transformed into grayscale images prior to the segmentation. Such a transformation may result in a loss of information. Moreover, conventional color image standards aim at reproducing human vision, and thus do not directly represent the physiology of plant leaves.

Among non-conventional imaging approaches, some approaches such as thermography or chlorophyll fluorescence depict the physiology of plant leaves through single channel images[14,15]. These single channel images may easily be segmented using automated thresholding procedures for the quantification of disease severity. Chlorophyll fluorescence analysis is a non-destructive technique that has been used for imaging plant pathogen interactions [16,17] and in particular to assess the resistance of plant to pathogen [18-23]. Indeed, symptoms result from the alteration of the tissues and many pathogens target the carbon metabolism and the photosynthetic apparatus [24-26]. Among all the chlorophyll fluorescence parameters that can be estimated, the maximum quantum yield of photosystem II (PSII) photochemistry ($F_v/F_m = (F_m-F_0)/F_m$) [27] is interesting for phenotyping disease severity as it is an indicator of plant stress[17,28]. $F_v/F_m$ is a parameter calculated from two measured fluorescence parameters, $F_0$(minimum fluorescence) and $F_m$ (maximum fluorescence). $F_v/F_m$ was reported to display a robust contrast between infected and healthy tissues [17,20,28,29]. Furthermore, healthy tissues were reported to yield $F_v/F_m$ values around 0.84 for numerous plant species [30,31]. When tissues are altered by biotic or abiotic stress, $F_v/F_m$ values decrease [32-34]. In many studies, mean $F_v/F_m$measurements were used to qualitatively discriminate between diseased and healthy leaves[23,33,35-37]. However, quantitative assessments of the total diseased area on leaves require each pixel to be classified as diseased or healthy. Thresholding based on $F_v/F_m$ values may allow the segmentation of diseased areas on the imaged leaves.

Common Bacterial Blight (CBB) of bean is caused by *Xanthomonas axonopodis* pv. *phaseoli* and *X. fuscans* subsp. *fuscans* (*Xff*). These pathogens are listed by the European and Mediterranean Plant Protection Organization [38] as quarantine pathogens as CBB is the most destructive bacterial disease of the common bean *Phaseolus vulgaris* resulting in up to 60% yield losses in favorable conditions [39]. On leaves, visible symptoms start usually four to seven days after infection with pinpoint water soaked areas that enlarge and eventually form necrotic tissues surrounded by a chlorotic halo. Attack may also result in leaflet wilting and in severe cases to defoliation [38]. Furthermore, the agents of CBB may also accomplish their whole cycle in the absence of visible symptoms [40]. As for other plant bacterial diseases, no efficient chemical treatment is allowed in the European Union, and control of the disease mainly involves the use of resistant bean cultivars.

No total monogenic resistance to the agents of CBB is known in *P. vulgaris*. However, quantitative trait loci conditioning resistance to CBB have been identified and quantitative resistances may be bred into commercial cultivars of bean [41]. Therefore quantification of the total diseased area on bean leaflets is needed to monitor the resistance level of novel bean lines to the agents of CBB during the selection process. In the present study, we developed procedures for the automated segmentation of $F_v/F_m$ images in order to quantify disease severity on plant leaflets in the pathosystem *P. vulgaris/Xff* CFBP4834-R. At first, we explored expert-defined thresholds after visual observations to discriminate in $F_v/F_m$ images areas corresponding to necrotic, wilted, impacted tissues, and healthy tissues. Second, we tested a segmentation approach based on modeling pixel-wise $F_v/F_m$-distributions as mixtures of Gaussian distributions, each distribution representing a different stage of the alteration of plant tissues, from strongly altered to healthy tissues. Finally, we developed a thresholding approach based on the probability of misclassification of a healthy pixel into the class of diseased pixels. Then, the segmented diseased areas can be modeled as mixtures of Gaussian distributions to discriminate various stages of alteration of plant tissues, from strongly to weakly altered tissues.

# RESULTS

## Datasets

Two datasets were used. The first one was used to setup the segmentation procedure for the quantification of symptoms by image analysis. It features images of leaflets of *P. vulgaris* cultivar (cv.) Flavert plants inoculated with the strain *Xff* CFBP4834-R or water. In this first dataset, the imaged leaflets were not detached from the plant to monitor the development of symptoms during 11 days after inoculation (dai). The same leaflets were imaged first at 1 dai then every day between 4 and 11 dai to monitor the development of the symptoms. For each leaflet, the symptomatic area was delimited either by: i) thresholding based on expert visual observations, ii), thresholding based on modeling pixel-wise $F_v/F_m$-distributions as mixtures of Gaussian distributions or iii) thresholding based on the probability of misclassification of a healthy pixel followed by a subsequent clustering of diseased pixels to describe the various stages of alteration of plant tissues.

A second dataset featured images of leaflets of five bean cultivars (cvs. Flavert, Michelet, Pike, Caprice and Wonder) inoculated with *Xff* CFBP4834-R. The image analysis procedure previously calibrated on the first dataset was applied to images belonging to this second dataset for evaluating the resistance of these commercial bean cultivars to

*Xff* CFBP4834-R. In this second dataset, leaflets were detached from inoculated plants just before imaging to ease and speed up the image acquisition.

## Thresholding Based on Expert Visual Observations

Expert-based thresholding consisted in the comparison of conventional color images with $F_v/F_m$ images of the same leaflet by trained raters to manually define the relevant thresholds to segment the $F_v/F_m$ images. The segmented $F_v/F_m$ images should visually reproduce the distribution of symptoms as visualized on conventional color images, i.e. the various segmented parts in $F_v/F_m$ images should co-localize with the various stages of the symptom development as observed by the eye of trained raters. On bean leaflets of cv. Flavert harboring symptoms of *Xff*CFBP4834-R, we could discriminate between necrotic tissues, wilted, impacted and healthy tissues. Water soaked symptoms displayed $F_v/F_m$ values similar to that of wilted tissues, therefore both are referred to as wilted tissues. Some tissues that did not harbor any visible symptoms displayed similar $F_v/F_m$ values as chlorotic tissues, therefore both are referred to as impacted tissues. Subsequently, three $F_v/F_m$ thresholds were determined to allow the automated segmentation of necrotic, wilted, impacted and healthy tissues (Figure 1).

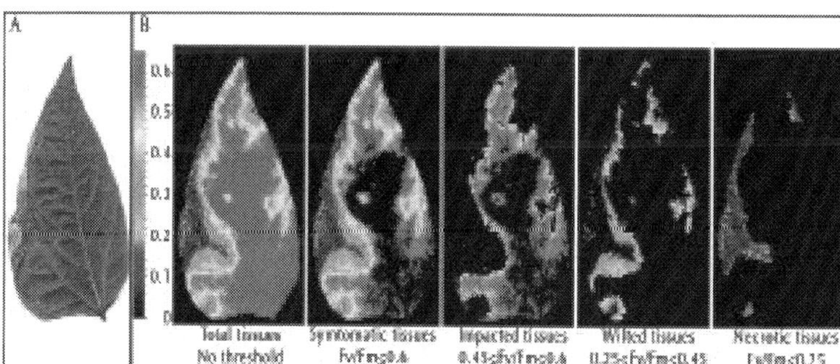


**Figure 1:** Expert-based thresholds allow the segmentation of various stages of the symptom development. Two weeks-old bean plants cv. Flavert were inoculated with either *Xff* CFBP4834-R ($1.10^6$ CFU $ml^{-1}$) or mock. This leaflet inoculated with *Xff* CFBP4834-R was sampled on bean *P.vulgaris* cv. Flavert

at 11 dai. A: visible image obtained by scanning. Necrosis is clearly visible on the left marge of the leaflet surrounded by wilted tissues. B: $F_v/F_m$ image obtained by chlorophyll fluorescence imaging. The three stages of the symptom development, i.e. necrotic, wilted and impacted tissues, were segmented respectively with the thresholds $0.25 \leq F_v/F_m$, $0.25 < F_v/F_m \leq 0.45$ and $0.45 < F_v/F_m \leq 0.6$. Black areas represent non-selected pixels with the threshold. After the segmentation step, the proportion of pixels in each segment may be quantified.

A training subset of images was used by the trained raters to define thresholds. The comparison of visible and $F_v/F_m$ images revealed that pixels displaying $F_v/F_m$ values inferior to 0.6 co-localized with a diseased area. $F_v/F_m$ values ranging from 0 to 0.25 co-localized with necrotic tissues, whereas $F_v/F_m$ values ranging from 0.25 to 0.45 co-localized with wilted tissues. $F_v/F_m$ values ranging from 0.45 to 0.6 corresponded to impacted tissues.

Expert-based thresholds were applied on all $F_v/F_m$ images to quantify each stage of the symptom development. Significant symptom development began at 7 dai on leaflets inoculated with *Xff*CFBP4834-R (*p-value* $< 0.01$). From 9 dai on, the symptoms were predominantly composed by wilted tissues (Figure 2A).

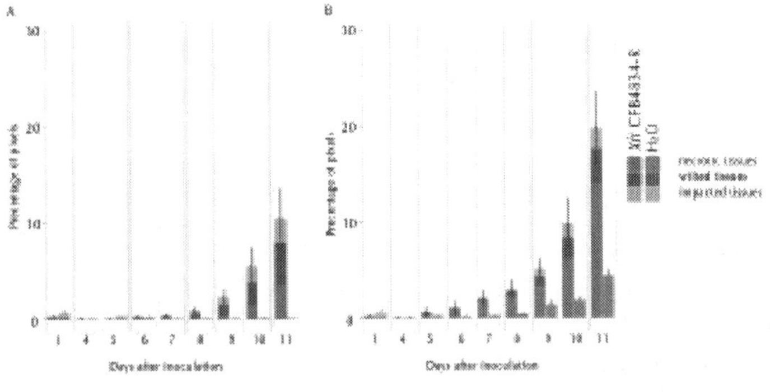

**Figure 2:** Evolution of the proportions of necrotic, wilted and impacted tissues on bean leaflets using expert-based thresholding. Two weeks-old bean plants cv. Flavert were inoculated with either *Xff* CFBP4834-R ($1.10^6$ CFU $ml^{-1}$) or mock. Observations were made on bean leaflets sampled at 1 dai, and everyday after the fourth dai. Percentages of diseased tissues and standard error of the mean were calculated for 20 leaflets per sampling day. The percentages do not include (A) or include (B) the estimation of the shrinking of the leaflet. The shrinking of the leaflet is attributed to necrotic tissues.

Moreover, as in this first dataset we monitored the evolution of symptoms on the same leaflets over time, we could assess shrinking of leaflets induced by the pathogen. Shrinking corresponds to the difference between the maximum size of the leaflet and its current size at the observation timepoint. Shrinking of leaflets may represent up to 10% of the total leaflet area (Figure 2). It mostly corresponds to necrotic tissues. However, some shrinking may also originate from natural aging of leaflets as shrinking was also detected on mock-inoculated leaflets.

## Thresholding Based on Modeling Pixel-wise $F_v$/$F_m$ Distributions

Abiotic stresses during the experiment may affect the physiological status of plants, which will in turn impact the $F_v/F_m$ values. In such cases, applying fixed thresholds may generate segmentation artifacts. To avoid such a drawback, mock-inoculated leaflets were used as internal daily controls to setup daily thresholds. Indeed, thresholds were defined daily to take into account the daily $F_v/F_m$ variations.

From the $F_v/F_m$ images of leaflets mock- or *Xff* CFBP4834-R-inoculated, we extracted the pixel-wise $F_v/F_m$-distributions. Analyzing these distributions may help defining appropriate non-overlapping clusters of $F_v/F_m$ values that represent various stages of alteration of plant tissues. In statistics, Gaussian mixture model (i.e. weighted sum of Gaussian distribution) is often used for clustering analysis. A clustering method proposed by Hennig [42] was tested in the present study to identify clusters of tissues according to pixel-wise $F_v/F_m$-distributions. Gaussian mixture model can be used to formalize the underlying heterogeneous distribution of $F_v/F_m$ values that are assumed to be composed of several clusters of pixels types, each cluster being modeled by a Gaussian distribution. From a practical point of view, Gaussian distributions are often too stiff to model true cluster shapes and Gaussian mixture modeling methods tend to select "larger optimal number of mixture components than what seems to be reasonable when looking to the data" [42]. Among the various methods proposed by Hennig [42] to solve this problem, we have tested the ridgeline unimodal method where clusters are merged if their fusion results in an unimodal Gaussian distribution [42,43].

In practice, using the R package MCLUST [44], the pixel-wise $F_v/F_m$-distributions were first fitted to Gaussian mixture models having from 1 to 4 components: one cluster should group pixels representing healthy tissues and three clusters should group pixels representing the various stages of the symptom development. The model best supported by the data is then determined using BIC criteria and lastly corrected by the ridgeline unimodal method [42,43]. For each pixel in the image, a probability of membership to each cluster is estimated. Clusters gathering less than 1% of the pixels were considered as not significant. Pixels initially attributed to these non-significant clusters were assigned according to their second better probability of membership.

On each image of mock-inoculated leaflets, pixel-wise $F_v/F_m$-distribution could be modeled as one single cluster per image. Predicted means of these clusters grouping pixels corresponding to healthy tissues may vary among images of mock-inoculated leaflets (e.g. from 0.71 to 0.85 at 4 dai). Hence, we calculated a confidence interval for predicted means of clusters grouping pixels corresponding to healthy tissues. Conversely, on diseased leaflets, $F_v/F_m$ values were modeled by mixtures involving more than one cluster. Among these, one cluster grouped pixels corresponding to healthy tissues, while additional clusters contained pixels corresponding to diseased areas. To discriminate between healthy and diseased tissues, a threshold based on the lower limit of the confidence interval previously defined on mock-inoculated samples was used. Subsequently, the quantification of pixels corresponding to diseased area allows the calculation of the proportion of diseased tissues on each leaflet. As expected, the proportion of diseased area on inoculated leaflets increased from 7 dai to the end of the experiment (Table 1).

**Table 1:** Quantification of the diseased tissues using the expert-, the model- and the probability-based thresholding approaches

| | | **1** | **4** | 5 | **6** | 7 | **8** | 9 | | |
|---|---|---|---|---|---|---|---|---|---|---|
| Expert-based thresholding | Necrotic tissues (%) | 0,02 | 0,00 | 0,00 | 0,02 | 0,03 | 0,08 | 0,39 | 1,46, | 3,44 |
| | Wilted tissues (%) | 0,06 | 0,01 | 0,01 | 0,06 | 0,19 | 0,49 | 1,02 | 2,15 | 4,41 |
| | Impacted tissues (%) | 0,16 | 0,05 | 0,06 | 0,13 | 0,20 | 0,37 | 0,90 | 1,74 | 2,56 |
| | Total diseased tissues (%) | 0,24 | 0,06 | 0,08 | 0,22 | 0,42 | 0,93 | 2,30 | 5,35 | 10,41 |
| Model-based thresholding | Total diseased tissues (%) | 0,00 | 5,26 | 5,26 | 23,87 | 42,96 | 5,64 | 35,62 | 49,01 | 30,39 |
| Probability-based thresholding | Strong alteration (%) | 0,01 | 0,00 | 0,00 | 0,00 | 0,00 | 0,00 | 0,00 | 1,35 | 2,70 |
| | Moderate alteration (%) | 0,08 | 0,01 | 0,03 | 0,11 | 0,28 | 0,70 | 1,85 | 2,52 | 6,67 |
| | Weak alteration (%) | 0,00 | 0,95 | 0,04 | 0,14 | 0,45 | 0,66 | 1,74 | 2,96 | 7,27 |
| | Total diseased tissues (%) | 0,09 | 0,95 | 0,07 | 0,25 | 0,73 | 1,36 | 3,59 | 6,83 | 16,64 |

Expert-based thresholding consists in defining $F_v/F_m$ thresholds that enable the selection of areas on $F_v/F_m$ images that match the various stages of the symptom development as observed by trained raters on conventional color images. Healthy, necrotic, wilted, or tissues impacted by the pathogen can be quantified.

Model-based thresholding consists in modeling the pixel-wise $F_v/F_m$-distributions extracted from each image by mixtures of Gaussian distributions. Such modeling results in the definition of clusters of pixels that correspond to various stages of the alteration of plant tissues. This step is based on the sole analysis of pixel-wise $F_v/F_m$-distributions and not on the visual observation of symptoms on conventional color images. Therefore, we use the terminology of strong, moderate and weak alteration, to emphasize that this classification is not based on visual observations, and does not necessarily correspond to the various stages of the symptom development as observed by trained raters. When applied directly without preliminary delimitation of the total diseased area, such a modeling over weights artifacts that can occur on healthy tissues, which results in a large overestimation of the proportion of diseased tissues.

Probability-based thresholds consists in the 500-quantile of the merged pixel-wise $F_v/F_m$-distributions of mock-inoculated samples. Each day of the experiment, the $F_v/F_m$probability-based threshold allows the splitting of pixels corresponding to healthy and diseased areas. Then whithin the diseased area only, pixel-wise $F_v/F_m$-distributions are modeled as mixtures of Gaussian distributions to quantify various stages of alteration of plant tissues.

Pixels were recolored according to the cluster they belong to (Figure 3). Such a model-based clustering allows the discrimination between various stages of alteration of the plant tissues. These various stages of alteration of plant tissues strictly depend on the structure of pixel-wise $F_v/F_m$-distributions and are thus independent of any *a priori* based on visual observation of symptoms (Figure 3).

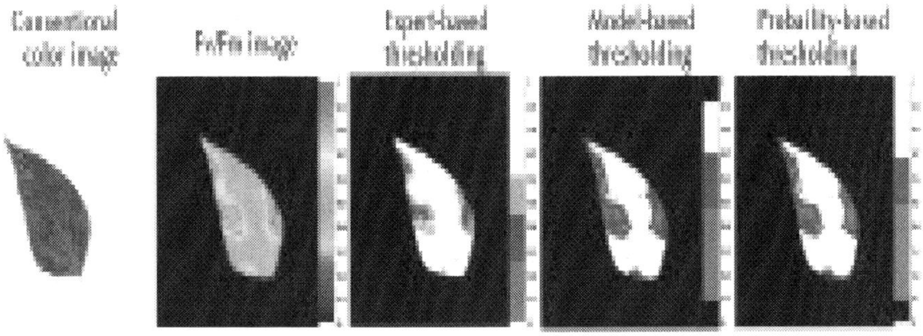

**Figure 3:** Mapping of the diseased areas segmented using the three thresholding approaches. Two weeks-old bean plants cv. Flavert were inoculated with either *Xff* CFBP4834-R (1.10$^6$ CFU ml$^{-1}$) or mock. This leaflet inoculated with-*Xff* CFBP4834-R was sampled on bean cv. Flavert at 11 dai. A: visible image obtained by scanning. B: $F_v/F_m$ image obtained by chlorophyll fluorescence imaging. C-E: Segmentation of the $F_v/F_m$ image for selection of the diseased area. C: Expert-based thresholds are defined after comparison with visual observations by trained raters. D: Model-based thresholds are defined by the clustering approach on the total surface af the leaflet. E: Probability-based thresholds are defined on the probability that a healthy pixel is misclassified with a specificity of 0.002. The healthy tissues are represented in white. The diseased tissues are colored in red, blue or green. Defioliation spots are represented in black.

Such an approach over-weights artifacts that can occur on non-symptomatic samples. Indeed, the rule for thresholding involves the mean of the predicted distribution, but not the own $F_v/F_m$ value of each pixel. Therefore, on a mock-inoculated leaflet, if the mean of the single predicted distribution is slightly under the threshold, the whole leaflet area will be considered as diseased tissues. Such a caveat may also occur for the predicted distribution grouping pixels of healthy areas on *Xff* CFBP4834-R-inoculated leaflets, resulting in a strong overestimation of the diseased areas (Table 1).

## Thresholding Based on the Probability of Misclassification of a Healthy Pixel

To solve the caveat of the overestimation of the diseased area on inoculated leaflets, we decided to normalize on mock-inoculated samples by classifying each pixel based on the probability of misclassification of a healthy pixel. Probability-based thresholds were defined on the pixel-wise $F_v/F_m$-distributions of mock-inoculated leaflets. Thresholds were defined daily to take into account the daily $F_v/F_m$ variations. Day by day, the pixel-wise $F_v/F_m$-distributions of all the mock-inoculated leaflets were merged. The resulting distribution thus represents all the possible values for a healthy pixel, including some abnormally low $F_v/F_m$ values. The $F_v/F_m$ values corresponding to the 100-quantile, 500-quantile or 1000-quantile, i.e. the $F_v/F_m$ values splitting 1/100, 1/500 and 1/1000 of the pixels of the distribution of the mock-inoculated leaflets, were used as thresholds splitting infected and healthy tissues. For each $F_v/F_m$ image, pixels were recolored according to these thresholds. The 1/1000 threshold was too stringent, as the segmented area does not contain the totality of the visible symptom. The 1/100 threshold was not stringent enough, as randomly distributed pixels were selected in addition to symptoms. Thus, we chose the 1/500 threshold as the totality of the symptom was segmented and no randomly distributed pixels were selected (data not shown). Using such a threshold the specificity of the approach is 0.002, i.e. there was a probability of 0.002 to misclassify a healthy pixel as diseased.

The probability-based thresholds are presented in Figure 4A by a vertical solid bar. Thresholds varied according to the day of the experiment (e.g. 0.467 at 1 dai, 0.689 at 7 dai and 0.722 at 11dai, Figure 4A) indicating that daily variations in the $F_v/F_m$ status of plants occurred during the experiment. For each leaflet, pixels exhibiting $F_v/F_m$ values lower than these probability-based thresholds were considered as diseased. Significant amounts of symptoms first arose at 7 dai (*p-value* $< 0.01$, Figure 4B, Table 1).

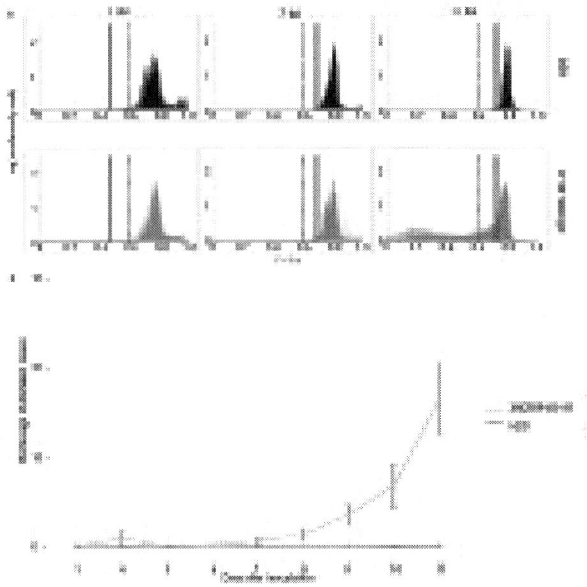

**Figure 4:** Daily thresholds for calculation of the proportion of diseased tissues with the probability-based approach. Two weeks-old bean plants cv. Flavert were inoculated with either *Xff* CFBP4834-R ($1.10^6$ CFU $ml^{-1}$) or mock. A: Pixel-wise $F_v/F_m$-distributions at 1, 7 and 11 dai of mock-inoculated leaflets (black) and *Xff* CFBP4834-R-inoculated leaflets (grey). The expert-based thresholds (dotted bars) are fixed over the whole experiment. Conversely, the probability-based thresholds (solid bars) may vary each day of the experiment, thereby taking into account daily physiological variations of plants. B: Evolution of the percentage of diseased areas on mock-inoculated (black curve) and *Xff*CFBP4834-R-inoculated (grey curve) leaflets calculated with probability-based thresholding approach.

In order to discriminate between various stages of alteration of plant tissues, the pixel-wise $F_v/F_m$-distributions corresponding to the diseased tissues were extracted. Using MCLUST [44], these distributions were first fitted to Gaussian mixture models having up to 3 components to match the various stages of alteration of plant tissues (Table 1). Mapping these clusters on the recolored $F_v/F_m$ images confirmed that they match various stages of the alteration of plant tissues, i.e. weak alteration, moderate alteration, strong alteration.

## Evaluation of the Resistance of Commercial Bean Cultivars to Xff Cfbp4834-R

Five bean cultivars (cvs. Flavert, Michelet, Pike, Wonder and Caprice) were inoculated with the strain *Xff* CFBP4834-R or water. Symptoms were quantified by our image analysis procedure. The Expert-based thresholding approach was not used as the thresholds were defined from a training set of images only on cv. Flavert but not on the other cultivars. Instead, the total amount of diseased areas was determined using probability-based thresholds. Then, using a clustering approach (that does not require training datasets), the various stages of the alteration of plant tissues were discriminated and quantified.

$F_v/F_m$ images were taken on detached leaflets at 7 and 11 dai and the amounts of diseased tissues on leaflets were calculated (Figure 5). During the experiment, the amount of diseased tissues increased for all the cultivars tested. However, differential behaviors among the various cultivars tested could be observed. At both 7 and 11 dai, cv. Flavert exhibited a significantly higher amount of symptoms than the other bean cultivars (*p-value* $< 0.05$), indicating that cv. Flavert is the most sensitive cultivar to *Xff* CFBP4834-R. On the contrary, the amounts of symptoms detected on cvs. Wonder and Caprice significantly differed from the mock-inoculated samples only at 11 dai. At 11 dai, cvs. Caprice and Wonder displayed the weakest total amount of symptoms among all cultivars tested, indicating that these cultivars were the most tolerant to *Xff* CFBP4834-R in this study. At both 7 dai and 11 dai, cvs. Michelet and Pike exhibited amounts of symptoms significantly higher than mock-inoculated samples (*p-value* $< 0.05$). At 11 dai the total amounts of symptom detected on cvs. Michelet and Pike were significantly higher than that detected on cvs. Caprice and Wonder. Therefore, our study revealed three levels of tolerance to *Xff* CFBP4834-R at 11 dai: cv. Flavert was sensitive, cvs. Michelet and Pike were partially tolerant, and cvs. Caprice and Wonder were tolerant.

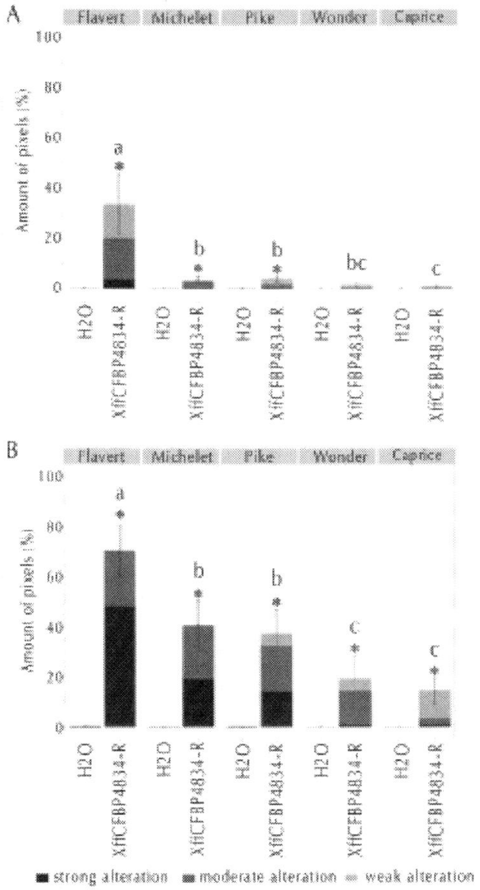


**Figure 5:** Quantification of the proportion of diseased tissues caused by *Xff* CFBP4834-R on five cultivars of bean. Bean leaflets of cultivars Flavert, Michelet, Pike, Wonder and Caprice were inoculated with *Xff*CFBP4834-R ($1.10^6$ CFU $ml^{-1}$) or mock. Observations were made at 7 (A), and 11 dai (B). On each leaflet diseased area was segmented using the probability-based thresholds. Means of percentages of diseased tissues and standard error of the mean were calculated from two repeats with 4 (7 dai) and 7 (11 dai) leaflets. Treatments denoted by different letters are significantly different (*p-value* < 0.01) based on the Mann–Whitney test. Asterisks mark significant differences between the mock-inoculated and CFBP4834-R-inoculated leaflets (*p-value* < 0.01) based on the Mann–Whitney test. Clusters representing the various stages of the alteration of plant tissues were determined by a clustering approach using MCLUST [44] on the diseased area only.

Then, the clustering approach was applied on the diseased areas to discriminate the various stages of the alteration of plant tissues i.e. weak alteration, moderate alteration and strong alteration. Therefore, an optimal number of three clusters was estimated. At 7 dai, symptoms on cv. Flavert, were essentially composed by moderately altered tissues. Some weakly altered tissues were also detected. At 11 dai, cv. Flavert exhibited a high amount of strongly altered tissues in comparison with the other cultivars. Symptoms segmented on cvs. Wonder and Caprice comprised weakly and moderately altered tissues. Cvs. Pike and Michelet presented an intermediary level of severity symptoms, i.e. moderately altered tissues and few strongly altered tissues.

# DISCUSSION

During the last decade, non-conventional imaging techniques such as chlorophyll fluorescence imaging were used for the study of the interactions between plant and pathogens [17]. Indeed, with chlorophyll fluorescence imaging, contrasts are enhanced compared to conventional color images, and depict more accurately the physiology of plant tissues [14,15]. Chlorophyll fluorescence imaging provides images that map on leaves the variations of single parameters associated to photosynthesis. Among the various fluorescence parameters, we monitored variations in the maximum quantum yield of photosystem II photochemistry ($F_v/F_m$). We used chlorophyll fluorescence imaging to map on bean (*P. vulgaris*) leaflets the areas altered by *Xff*CFBP4834-R. It must be pointed out that most stresses that decrease leaf health will affect photosynthesis. Therefore we first checked that visible symptoms of CBB on bean leaflets co-localized with decreased values of the $F_v/F_m$ parameter. Other parameters such as $F_v/F_0$ or $F_0/F_m$ were shown to yield a high contrast between healthy tissues and tissues affected with various pathogens [20,45,46]. However, we did not treat these parameters in this study as they do not have a clear physiological significance [20].

Images based on the single $F_v/F_m$ parameter are easier to segment by thresholding approaches than conventional color images, thereby easing the image analysis process. In such a context, we developed an automated thresholding procedure to select pixels corresponding to symptoms. The respective amounts of pixels corresponding to diseased or healthy areas can then be quantified to assess the disease severity on inoculated plants. Even though the decrease in $F_v/F_m$ values due to the pathogen attack is now well documented [32,33], only few studies developed approaches for the quantification of the diseased area on leaves. Most of the studies using the $F_v/F_m$ parameter in plant pathology are based on the mean $F_v/F_m$ value over the whole image[23,33,35-37,47]. The mean $F_v/F_m$ value may qualitatively discriminate between healthy and diseased leaves, but does not quantify the amount of diseased tissues [23,33,35-37,47]. Only few studies attempted to analyze the pixel-wise $F_v/F_m$-distribution to discriminate between healthy and diseased organs using a threshold of $F_v/F_m$. For example, in the case of *Fusarium culmorum*, wheat ears were considered as infected when pixels displaying a $F_v/F_m$ value lower than 0.3 could be observed in the image [18].

In the present study, we investigated thresholding approaches for the quantification of the diseased area on infected leaves. At first, trained raters compared a subset of visible images and $F_v/F_m$ images of bean leaves of cv. Flavert inoculated with *Xff* CFBP4834-R. We could associate non-overlapping clusters of $F_v/F_m$ values to each stage of the symptom development (necrotic, wilted, impacted and healthy tissues) caused by *Xff* CFBP4834-R on *P. vulgaris* cv. Flavert. Subsequently $F_v/F_m$ thresholds could be defined to discriminate between the various stages of the symptom development. Counting the pixels associated to each $F_v/F_m$ cluster enables the quantification of the leaf area corresponding to each stage of the symptom development on cv. Flavert. Defining non-overlapping clusters of $F_v/F_m$ values to segment symptomatic areas can also be performed on a broad range of plant species to quantify areas affected by biotic or abiotic stresses. For example, on grapevine affected by lime-induced iron chlorosis, chlorotic areas displayed lower $F_v/F_m$ values than healthy tissues [48].

Interestingly, lower $F_v/F_m$ values may be observed on tissues located in the margin of symptomatic areas, but that do not display any visible symptoms. These areas evolve into symptoms over time. Therefore, as previously observed on *Arabidopsis thaliana* or *Nicotiana benthamiana* inoculated with *Pseudomonas syringae*, pre-symptomatic areas may also be phenotyped using $F_v/F_m$[33,36]. $F_v/F_m$ values in these tissues may not differ from those observed in chlorotic tissue and both chlorotic and pre-symptomatic tissues were grouped into impacted tissues in the present study. The decrease of the $F_v/F_m$ values in pre-symptomatic areas is not fully understood. Indeed, neither these areas are yet colonized by bacteria, nor can be observed increased levels of ammonia or a restricting water movement [36].

However, non-overlapping clusters matching the various stages of the symptom development should be defined by trained raters in each pathosystem studied. Indeed, $F_v/F_m$ clusters defined on Flavert do not match the visual observation on other cultivars. For example, the cluster corresponding to visually chlorotic tissues on cv. Michelet overlaps with that corresponding to necrosis on cv. Flavert. Therefore fixed thresholds defined on a correspondence with visual observations by trained raters are valuable only within a single cultivar, and cannot be extrapolated to other cultivars. Using an expert-based thresholding approach on other cultivars needs a calibration step on a training set of pictures. Such a need is a limitation for this thresholding approach in the perspective of high throughput phenotyping, as visual assessment is time consuming.

Therefore, we aimed at defining $F_v/F_m$ thresholds that could be extrapolated to any cultivar or plant species. As a decrease of $F_v/F_m$ depicts the alteration of PSII, $F_v/F_m$ threshold can be defined independently of visual observations. Clusters of $F_v/F_m$ would depict objective stages of alteration of plant tissues. Moreover, the $F_v/F_m$ parameter may be impacted by the physiological status of plants [49,50] or abiotic stresses [16,34,51]. Fixed thresholds may therefore bias the quantification of the diseased area on leaves. Defining the thresholds on control plants for each experimental round helps avoid such a bias.

To avoid the use of fixed thresholds, we normalized our segmentation on mock-inoculated plants. We defined the threshold as the $F_v/F_m$ value under which a healthy pixel only has a probability of 0.002 to be misclassified. Such a thresholding does not allow the discrimination of various stages of the symptom development. Therefore, within the diseased area, the pixel-wise $F_v/F_m$-distribution was modeled as a mixture of predicted Gaussian distributions. Such a modeling is largely used for image analysis in medical sciences [52,53]. As well, in plant sciences such a modeling was recently applied to the automated recognition of individual Arabidopsis rosettes, in order to monitor independently the growth of each plant in the image [54]. In the present study, the clustering of pixels according to Gaussian distributions aims at describing the various stages of alteration of plant tissues without any *a priori* based on visual observations. For each $F_v/F_m$ image, a mixture of Gaussian is fitted independently, and such an approach does not need any calibration set of images. Based on these Gaussian mixture distributions, we could define non-overlapping clusters of pixels displaying similar $F_v/F_m$ values, corresponding to the various stages of alteration of plant tissues.

Pathogen attack may also result in dwarfing or shrinking of leaves. Such a phenotype is rarely quantified [9,12], but the use of non-destructive image analysis approaches may help solve such a caveat. In the present study, we monitored the size of leaflets over time. We considered the maximum size of each leaflet as a reference and the size decrease compared to this reference was considered as shrinking. Such a phenotyping is difficult to assess by visual observation only. A similar approach was used to analyze the leaf area impacted by herbivory [9]. Other approaches were proposed to evaluate the leaf deformation, for example using a sphericity index [12,55]. However, using such an index does not allow the quantification of the leaf area impacted by the shrinking.

To test the applicability of our segmentation approach for the evaluation of plant resistance, we quantified the symptoms caused by *Xff* CFBP4834-R on five commercial cultivars of bean (cvs. Flavert, Michelet, Pike, Wonder and Caprice). When looking at the total amounts of symptoms, the cv. Flavert appeared to be the most sensitive to *Xff* CFBP4834-R. The cvs. Caprice and Wonder exhibit few symptoms and can be considered as tolerant. The cvs. Michelet and Pike are impacted to an intermediary extent. On top of the amount of symptomatic tissues, selection for resistance may also focus on the stage of development of the symptom. Indeed, it may be of interest for breeders to notice that cvs. Caprice and Wonder exhibit different symptoms topologies, even though they displayed similar total amounts of symptomatic tissues.

Finally, in order to select for quantitative plant resistance to pathogens, high throughput procedures aiming at precisely quantifying disease severity need to be developed. Robotic imaging procedures can increase the number of images taken [56] but few automatic analysis of chlorophyll fluorescence images procedures are available. The procedure presented in this study was automated under R and the R script is available at http://lisa.univ-angers.fr/PHENOTIC/telechargements.html *webcite*. Running our procedure on the 1080 images of our dataset, two minutes only are needed for the Expert- and Probability-based thresholding analyses. The use of MCLUST [44] to discriminate various stages of alteration of plant tissues increases up to one hour the calculation time, which remains much faster than rating disease severity by visual observations.

# CONCLUSIONS

In this paper, we described new procedures to quantify the impact of a pathogen on a plant, easy to automate, objective and accurate.

The expert-based thresholding approach aims at reproducing the visual observations. Such an approach allows the accurate quantification of the various stages of the symptoms development but needs to be calibrated by trained raters on each pathosystem. In contrast, a probability-based thresholding approach may accurately discriminate between healthy and diseased tissues. Within the diseased area, a clustering approach may accurately describe the various stages of alteration of plant tissues. This latter segmentation approach is expert-independent and is normalized on mock-inoculated plants at each day of the experiment. Moreover, the probability-based thresholding approach may allow the phenotyping of pre-symptomatic areas, which cannot be achieved by calibrating thresholds on visual observations by trained raters. A clustering approach applied on the diseased areas allows the quantification of each stage of the alteration of plant tissues. The segmentation approach developed in this study was automated using R, and the script is available at http://lisa.univ-angers.fr/PHENOTIC/telechargements.html *webcite*.

Such a development of automated segmentation approach speeds up the assessment of disease severity on plants. It may reveal a significant improvement for high throughput testing of the plant resistance to pathogens during breeding.

# METHODS

## Biological Material

The bacterial strain *Xff* CFBP4834-R used in this study was obtained from the French Collection of Bacteria associated to Plants (CFBP, IRHS, Angers, France, http://www.angers.inra.fr/cfbp/*webcite*, accession n°4885). The strain was grown at 28°C in 10% TSA medium (tryptone at 1.7 g/L, soybean peptone at 0.3 g/L, glucose at 0.25 g/L, NaCl at 0.5 g/L, $K_2HPO_4$ at 0.5 g/L, agar at 15 g/L, pH 7.2).

Beans were individually seeded in plastic pots (7 × 7 × 8 cm) containing prewetted compost (NEUHAUS HUMINSUBSTRAT *N4*, NFU 44–551). Plants were grown in a controlled climatic room at 23°C/20°C (day/night) with a photoperiod of 16 h. Plants were watered three times per week and supplemented with N-P-K (18:14:18) at 0.3 g/liter once a week.

Two sets of plants were used in this study. The first set was used to setup approaches for quantification of symptoms by image analysis and was composed by forty plants of bean Flavert. This experiment was performed three times. A second set was used to evaluate the resistance of commercial bean cultivars to *Xff* CFBP4834-R and was composed by twenty-eight plants of five cultivars of bean obtained from Vilmorin (La Ménitré, France): Flavert, Caprice, Michelet, Pike and Wonder. Twenty-eight plants of each cultivar were used. This experimentation was repeated twice.

## Pathogenicity Assay

Bacterial suspensions calibrated at $1.10^8$ CFU.ml$^{-1}$ were made by harvesting bacterial cells from agar plates and suspending them in sterile distilled water. The inoculations were made at the trifoliate step by deeping half of batch of plants during 30 seconds in the diluted bacterial suspension to $1.10^6$ CFU.ml$^{-1}$. The other half served as control plants and was deeped in water. The first set of plants was incubated at 28°C/25°C (day/night) with a photoperiod of 16 h during 11 days and under high (70%) relative humidity. The second set of plant was incubated at 28°C/25°C (day/night) with a photoperiod of 16 h during 11 days and under high (95%) relative humidity. Plant inoculations were carried out under quarantine at UMR1345 IRHS, Centre INRA, Beaucouzé, France.

## Technical Setup and Image Acquisition

The PSI Open FluorCam FC 800-O (PSI, Brno, Czech Republic) was used to capture chlorophyll fluorescence images and to estimate the maximum quantum yield of PSII ($F_v/F_m$) of inoculated and control leaflets. The system sensor is a CCD camera with a pixel resolution of 512 by 512 and a 12-bit dynamic. The system includes 4 LED panels divided to 2 pairs. One pair provides an orange actinic light with a wavelength of around 618 nm, with an intensity that can vary from 200 to 400 µmol/m$^2$/s. It provides a 2s pulse that allows the measurement of the initial fluorescent state ($F_0$). The other pair provides a saturating pulse during 1s in blue wavelength, typically 455 nm, with an intensity of up to 3000 µmol/m$^2$/s. The saturating pulse allows collecting of the maximum fluorescence ($F_m$). Fluorescence chlorophyll imaging was used in a dark adapted mode after a dark period of 45 min [28] to produce maps with the fluorescent quantum efficiency $F_v/F_m = (F_m - F_0)/F_m$. For all image acquisitions, the observed leaflet is maintained horizontally for this sub-section. A kinetic analysis was performed: image acquisitions were done at 1 dai, and everyday after the fourth dai. The experiment was repeated three times, and in total 1080 images were collected. Another analysis consisted in collecting images of detached leaflets at 1, 7 and 11 dai, in total 280 images were collected. False color images representing $F_v/F_m$ values of the pixels and pixel-wise $F_v/F_m$-distributions were extracted for each leaflet. For a subset of leaflets, visible images were also taken with a digital camera to compare visual symptoms and $F_v/F_m$ values.

## Shrinking of the Leaflets

The shrinking of the leaflets was calculated when a kinetic analysis was performed on each leaflet. The shrinking corresponded to the difference between the maximum size of the leaflet during the experiment and the current size of the leaflet. The shrinking of the leaflet was added to the class presenting the lower $F_v/F_m$ values. The amounts of the other stages of the symptom development were calculated using the maximum size of the leaflet.

## Thresholding Based on Expert Visual Observations

The expert-based thresholding approach consisted in the determination of values characterizing the diseased tissues. Three stages of the symptom development were discriminated: necrotic, wilted and impacted tissues. Conventional color images and $F_v/F_m$ images were manually compared by trained raters to determine clusters of $F_v/F_m$ values matching with each stage of the symptom development. On the conventional color image, the various stages of the symptom development were manually delimited and the delimitations were superimposed on the $F_v/F_m$ image. The maximum and minimum $F_v/F_m$ values of each area that co-localized with each stage of the symptom development were determined. These values corresponded to the expert-based thresholds. The amounts of pixels contained in each subgroup were calculated.

## Thresholding Based on Modeling Pixel-Wise $F_v/F_m$ Distributions

The R package MCLUST [44] was run to select the number of Gaussian distributions that compose the Gaussian mixture model best supported by the data and to estimate the mean, variance and weight in the mixture distribution of each cluster. For each pixel, a probability of membership to each cluster is also returned. Gaussian mixture models including from 1 to 4 clusters with unequal variance were fitted to each leaflet pixel-wise $F_v/F_m$-distribution. The clusters representing less than 1% of the pixel-wise $F_v/F_m$-distribution were suppressed as they were considered as artifactual clusters. Pixels initially attributed to these non-significant clusters were assigned according to their second better probability of membership. The clusters of the Gaussian mixture model selected were then merged if their fusion was unimodal according to the ridgeline unimodal method implemented in the R package fpc [42,43]. A threshold was calculated for each dai to discriminate between distributions characterizing the healthy tissues and distributions characterizing the diseased tissues. The means of the clusters found on mock-inoculated samples were averaged. Then, the confidence interval was calculated. A cluster displaying a mean inferior to the threshold was considered as characterizing diseased tissues.

## Thresholding Based on the Probability of Misclassification of a Healthy Pixel

The probability based-thresholds were built based a pixel-wise $F_v/F_m$-distribution resulting from the merging of the pixel-wise $F_v/F_m$-distributions of all the mock-inoculated leaflets. The thresholds corresponded to the 100-quantile, 500-quantile or 1000-quantile, i.e. the $F_v/F_m$ values splitting 1/100, 1/500 and 1/1000 of the pixels of the distribution. The thresholds were determined for each dai and each cultivar to take into consideration the possible daily variations and the differences between cultivars. A pixel displaying a $F_v/F_m$ value inferior to the threshold was considered as diseased. The total diseased proportion was then calculated for each leaflet. Then, a clustering method using the R package MCLUST [44] was performed on the diseased area detected by the Probability-based thresholds to segment the diseased area according to three stages of alteration of plant tissues. Gaussian mixture models including from 1 to 3 clusters with unequal variance were fitted to the pixel-wise $F_v/F_m$-distribution of each leaflet. In order to classify the clusters into the various stages of alteration of plant tissues, a second clustering procedure with an optimal number of clusters between 1 and 3 was performed on the means of all the clusters. The clusters were then classified according to their membership to the clusters representing the stages of alteration of plant tissues.

## Visualization of the Diseased Tissues and Statistical Test

For the three thresholding steps, the stages of the symptom development were colored using the R package EBImage [57]. The pixels were colored according to the various thresholds.

Mann–Whitney test [58] was performed to compare amounts of diseased tissues between mock-inoculated and *Xff* CFBP4834-R-inoculated leaflets and between the various cultivars of bean.

# ACKNOWLEDGEMENTS

This work was supported by Angers Loire Métropole, Conseil Général Département de Maine-et-Loire and Région Pays de la Loire (Phenotic) and by Institut National de la Recherche Agronomique and Région Pays de la Loire. We are grateful to Valérie Grimault at Station Nationale Essais de Semences, Valérie Caffier, Carolyne Dürr and Philippe Simoneau (IRHS, Angers, France) and Laurent Noël (LIPM, Toulouse, France) for helpful discussions. We thank Marion Le Saux, Perrine Portier, Martial Briand and Géraldine Taghouti at French Collection of Plant-associated Bacteria (CFBP) for providing the bacterial strain. We wish to thank Elise Vandoeuvre at Vilmorin for providing the seeds. We thank Installations Expérimentales Mutualisées (INEM) for growing the plants.

# REFERENCES

1. Brun H, Chèvre A-M, Fitt BDL, Powers S, Besnard A-L, Ermel M, Huteau V, Marquer B, Eber F, Renard M, Andrivon D: Quantitative resistance increases the durability of qualitative resistance to Leptosphaeria maculans in Brassica napus. *New Phytol* 2010, 185:285-299.
2. Palloix A, Ayme V, Moury B: Durability of plant major resistance genes to pathogens depends on the genetic background, experimental evidence and consequences for breeding strategies. *New Phytol* 2009, 183:190-199.
3. Quenouille J, Montarry J, Palloix A, Moury B: Farther, slower, stronger: how the plant genetic background protects a major resistance gene from breakdown. *Mol Plant Pathol* 2013, 14:109-118.
4. Poland JA, Nelson RJ: In the eye of the beholder: the effect of rater variability and different rating scales on QTL mapping. *Phytopathology* 2011, 101:290-298.

5. Bock CH, Parker PE, Cook AZ, Gottwald TR: Visual rating and the Use of image analysis for assessing different symptoms of citrus canker on grapefruit leaves. *Plant Disease* 2001, 92:412-424.
6. Bock CH, Poole GH, Parker PE, Gottwald TR: Plant disease severity estimated visually, by digital photography and image analysis, and by hyperspectral imaging. *Crit Rev Plant Sci* 2010, 29:59-107.
7. Sherwood RT, Berg CC, Hoover MR, Zeiders KE: Illusions in visual assessment of Stagonospora leaf spot of Orchardgrass. *Phytopathology* 1983, 73:173-177.
8. Camargo A, Smith JS: An image-processing based algorithm to automatically identify plant disease visual symptoms. *Biosystems Engineer* 2009, 102:9-21.
9. Green JM, Appel H, Rehrig EM, Harnsomburana J, Chang J-F, Balint-Kurti P, Shyu C-R:PhenoPhyte: a flexible affordable method to quantify 2D phenotypes from imagery. *Plant Methods* 2012, 8:45.
10. Xie W, Yu K, Pauls KP, Navabi A: Application of image analysis in studies of quantitative disease resistance, exemplified using common bacterial blight-common bean pathosystem. *Phytopathology* 2012, 102:434-442.
11. Wijekoon CP, Goodwin PH, Hsiang T: Quantifying fungal infection of plant leaves by digital image analysis using Scion Image software. *J Microbiol Methods* 2008, 74:94-101.
12. Contreras-Medina LM, Osornio-Rios RA, Torres-Pacheco I, Romero-Troncoso RDJ, Guevara-González RG, Millan-Almaraz JR: Smart sensor for real-time quantification of common symptoms present in unhealthy plants. *Sensors (Basel, Switzerland)* 2012, 12:784-805.
13. Thilagamani S, Shanthi N: A survey on image segmentation through clustering. *Int J Res Rev Info Sci* 2011, 1:16-19.
14. Chaerle L, Van Der Straeten D: Seeing is believing: imaging techniques to monitor plant health. *Biochim Biophys Acta* 2001, 1519:153-166.
15. Sankaran S, Mishra A, Ehsani R, Davis C: A review of advanced techniques for detecting plant diseases. *Computers Electro Agri* 2010, 72:1-13.

16. Gorbe E, Calatayud A: Applications of chlorophyll fluorescence imaging technique in horticultural research: A review. *Sci Hortic* 2012, 138:24-35.
17. Rolfe SA, Scholes JD: Chlorophyll fluorescence imaging of plant–pathogen interactions. *Protoplasma* 2010, 247:163-175.
18. Bauriegel E, Giebel A, Herppich WB: Hyperspectral and chlorophyll fluorescence imaging to analyse the impact of Fusarium culmorum on the photosynthetic integrity of infected wheat ears. *Sensors (Basel, Switzerland)* 2011, 11:3765-3779.
19. Belin É, Rousseau D, Boureau T, Caffier V: Thermography versus chlorophyll fluorescence imaging for detection and quantification of apple scab. *Computers Electron Agri* 2013, 90:159-163.
20. Berger S, Benediktyová Z, Matouš K, Bonfig KB, Mueller MJ, Nedbal L, Roitsch T:Visualization of dynamics of plant-pathogen interaction by novel combination of chlorophyll fluorescence imaging and statistical analysis: differential effects of virulent and avirulent strains of P. syringae and of oxylipins on A. thaliana. *J Exp Botany* 2007, 58:797-806.
21. Chaerle L, Hagenbeek D, De Bruyne E, Valcke R, Van Der Straeten D: Thermal and chlorophyll-fluorescence imaging distinguish plant-pathogen interactions at an early stage. *Plant Cell Physiol* 2004, 45:887-896. PubMed Abstract | Publisher Full Text
22. Pérez-Bueno ML, Ciscato M, VandeVen M, García-Luque I, Valcke R, Barón M: Imaging viral infection: studies on Nicotiana benthamiana plants infected with the pepper mild mottle tobamovirus. *Photosynthesis Res* 2006, 90:111-123.
23. Pineda M, Olejníčková J, Cséfalvay L, Barón M: Tracking viral movement in plants by means of chlorophyll fluorescence imaging. *J Plant Physiol* 2011, 168:2035-2040.
24. Bilgin DD, Zavala JA, Zhu J, Clough SJ, Ort DR, De Lucia EH: Biotic stress globally downregulates photosynthesis genes. *Plant Cell Environ* 2010, 33:1597-1613.
25. Jelenska J, Yao N, Vinatzer BA, Wright CM, Brodsky JL, Greenberg JT: A J domain virulence effector of Pseudomonas syringae remodels host chloroplasts and suppresses defenses. *CB* 2007, 17:499-508.

26. Gürlebeck D, Jahn S, Gürlebeck N, Szczesny R, Szurek B, Hahn S, Hause G, Bonas U:Visualization of novel virulence activities of the Xanthomonas type. *Mol Plant Pathol* 2009, 10:175-188.
27. Butler WL: Energy distribution in the photochemical apparatus of photosynthesis. *Ann Rev Plant Physiol* 1978, 29:345-378.
28. Baker NR: Chlorophyll fluorescence: a probe of photosynthesis in vivo. *Ann Rev Plant Biol* 2008, 59:89-113.
29. Matouš K, Benediktyova Z, Berger S, Roitsch T, Nedbal L: Case study of combinatorial imaging: what protocol and what chlorophyll fluorescence image to use when visualizing infection of Arabidopsis thaliana by pseudomonas syringae? *Photosynthesis Res* 2006, 243:243-253.
30. Björkman O, Demmig B: Planta characteristics at 77 K among vascular plants of diverse origins. *Planta* 1987, 170:489-504.
31. Maxwell K, Johnson GN: Chlorophyll fluorescence - a practical guide. *J Exp Bot* 2000, 51:659-668.
32. Kuckenberg J, Tartachnyk I, Noga G: Temporal and spatial changes of chlorophyll fluorescence as a basis for early and precise detection of leaf rust and powdery mildew infections in wheat leaves. *Precision Agri* 2008, 10:34-44.
33. Bonfig KB, Schreiber U, Gabler A, Roitsch T, Berger S: Infection with virulent and avirulent P. syringae strains differentially affects photosynthesis and sink metabolism in Arabidopsis leaves. *Planta* 2006, 225:1-12.
34. Woo NS, Badger MR, Pogson BJ: A rapid, non-invasive procedure for quantitative assessment of drought survival using chlorophyll fluorescence. *Plant Methods* 2008, 4:27.
35. Muthuchelian K, Porta NLA, Bertamini M, Nedunchezhian N: Cypress canker induced inhibition of photosynthesis in field grown cypress (Cupressus sempervirens L.) needles. *Physiol Mol Plant Pathol* 2005, 67:33-39.
36. Iqbal MJ, Goodwin PH, Leonardos ED, Grodzinski B: Spatial and temporal changes in chlorophyll fluorescence images of Nicotiana benthamiana leaves following inoculation with Pseudomonas syringae pv. tabaci. *Plant Pathol* 2012, 61:1052-1062.

37. Zou J, Rodriguez-Zas S, Aldea M, Li M, Zhu J, Gonzalez DO, Vodkin LO, DeLucia E, Clough SJ:Expression profiling soybean response to Pseudomonas syringae reveals new defense-related genes and rapid HR-specific downregulation of photosynthesis. *MPMI* 2005, 18:1161-1174.

38. *European and Mediterranean Plant Protection Organization.*

39. Broughton WJ, Hern G, Blair M, Beebe S, Gepts P, Vanderleyden J: Beans (Phaseolus spp.) – model food legumes. *Plant Soil* 2003, 252:55-128.

40. Darrasse A, Bureau C, Samson R, Morris CE, Jacques M-A: Contamination of bean seeds by Xanthomonas axonopodis pv. phaseoli associated with low bacterial densities in the phyllosphere under field and greenhouse conditions. *Eur J Plant Pathol* 2007, 119:203-215.

41. Duncan RW, Singh SP, Gilbertson RL: Interaction of common bacterial blight bacteria with disease resistance quantitative trait Loci in common bean. *Phytopathology* 2011, 101:425-435.

42. Hennig C: Methods for merging Gaussian mixture components. *Adv Data Anal Classification* 2010, 4:3-34.

43. Ray S, Lindsay BG: The topography of multivariate normal mixtures. *Ann Stat* 2005, 33:2042-2065.

44. Fraley C, Raftery AE: MCLUST version 3 for R: normal mixture modeling and technical report No. 504. *Department* 2006, 1-57.

45. Nedbal L, Soukupová J, Whitmarsh J, Trtilek M: Postharvest imaging of chlorophyll fluorescence from lemons can be used to predict fruit quality. *Photosynthetica* 2000, 38:571-579.

46. Soukupova J, Smatanova S, Nedbal L, Jegorov A: Plant response to destruxins visualized by imaging of chlorophyll fluorescence. *Physiol Plant* 2003, 118:399-405.

47. Mandal K, Saravanan R, Maiti S, Kothari IL, Plants A, Vidyanagar V: Effect of downy mildew disease on photosynthesis and chlorophyll fluorescence in Plantago. *J Plant Dis Protect* 2009, 116:164-168.

48. Bavaresco L, Bertamini M, Iacono F: Lime-induced chlorosis and physio- logical responses in grapevine (Vitis vinifera L . cv . Pinot blanc) leaves. *Vitis* 2006, 45:45-46.

49. Aranda I, Castro L, Alía R, Pardos JA, Gil L: Low temperature during winter elicits differential responses among populations of the Mediterranean evergreen cork oak (Quercus suber). *Tree Physiol* 2005, 25:1085-1090.
50. Ogaya R, Peñuelas J: Comparative field study of Quercus ilex and Phillyrea latifolia: photosynthetic response to experimental drought conditions. *Environment Exp Bot* 2003, 50:137-148.
51. Jonaviciene K, Studer B, Asp T, Jensen LB, Paplauskienσ V, Lazauskas S: Identification of genes involved in a water stress response in timothy and mapping of orthologous loci in perennial ryegrass. *Biologia Plantarum* 2012, 56:473-483.
52. Wehrens R, Simonetti AW, Buydens LMC: Mixture modelling of medical magnetic resonance data. *J Chem* 2002, 16:274-282.
53. Li Q, Fraley C, Bumgarner RE, Yeung KY, Raftery AE: Donuts, scratches and blanks: robust model-based segmentation of microarray images. *Bioinformatics (Oxford, England)* 2005, 21:2875-2882.
54. De Vylder J, Vandenbussche F, Hu Y, Philips W, Van Der Straeten D: Rosette tracker: an open source image analysis tool for automatic quantification of genotype effects. *Plant Physiol* 2012, 160:1149-1159.
55. Pratt WK: *Digital Image Processing: PIKS Inside*. Third edition. New York, NY, USA: John Wiley; 2001:623-650.
56. Harbinson J, Prinzenberg AE, Kruijer W, Aarts MGM: High throughput screening with chlorophyll fluorescence imaging and its use in crop improvement. *Curr Opin Biotechnol* 2012, 23:221-226.
57. Pau G, Fuchs F, Sklyar O, Boutros M, Huber W: EBImage - an R package for image processing with applications to cellular phenotypes. *Bioinformatics (Oxford, England)* 2010, 26:979-981.
58. Sokal RR, Rohlf FJ: *Biometry: the principles and practice of statistics in biological research*. 4th edition. New York: W. H. Freeman and Co; 2012:937.

# Citations

## CHAPTER 1

Victoria V. Roshchina, "Vital Autofluorescence: Application to the Study of Plant Living Cells," International Journal of Spectroscopy, vol. 2012, Article ID 124672, 14 pages, 2012. doi:10.1155/2012/124672

## CHAPTER 2

Lucia Guidi and Elena Degl'Innocenti (2011). Imaging of Chlorophyll a Fluorescence: A Tool to Study Abiotic Stress in Plants, Abiotic Stress in Plants - Mechanisms and Adaptations, Prof. Arun Shanker (Ed.), ISBN: 978-953-307-394-1, InTech, DOI: 10.5772/22281.

# CHAPTER 3

Lobet et al.: An online database for plant image analysis software tools. Plant Methods 2013 9:38. doi:10.1186/1746-4811-9-38

# CHAPTER 4

Wang et al.: Estimating rice chlorophyll content and leaf nitrogen concentration with a digital still color camera under natural light. Plant Methods 2014 10:36. doi: 10.1186/1746-4811-10-36.

# CHAPTER 5

Zhou et al.: CalloseMeasurer: a novel software solution to measure callose deposition and recognise spreading callose patterns. Plant Methods 2012 8:49. doi: 10.1186/1746-4811-8-49.

# CHAPTER 6

0Wang et al.: Delivering high-resolution landmarks using inkjet micropatterning for spatial monitoring of leaf expansion. Plant Methods 2011 7:1. doi:10.1186/1746-4811-7-1

# CHAPTER 7

Ibáñez et al., Rapid metabolic profiling of Nicotiana tabacum defence responses against Phytophthora nicotianae using direct infrared laser desorption ionization mass spectrometry and principal component analysis Plant Methods 2010, 6:14. doi: 10.1186/1746-4811-6-14.

# CHAPTER 8

Mishra et al.: Chlorophyll fluorescence emission can screen cold tolerance of cold acclimated Arabidopsis thaliana accessions. Plant Methods 2014 10:38. doi:10.1186/1746-4811-10-38.

# CHAPTER 9

Voorend et al.: LEAF-E: a tool to analyze grass leaf growth using function fitting. Plant Methods 2014 10:37. doi:10.1186/1746-4811-10-37

# CHAPTER 10

Dean et al.: Use of diffusion magnetic resonance imaging to correlate the developmental changes in grape berry tissue structure with water diffusion patterns. Plant Methods 2014 10:35. doi:10.1186/1746-4811-10-35

# CHAPTER 11

Rousseau et al.: High throughput quantitative phenotyping of plant resistance using chlorophyll fluorescence image analysis. Plant Methods 2013 9:17. doi:10.1186/1746-4811-9-17

# Index